Geeking Gallium Gaffs & Gimmicks

The Disappearing Spoon Revitalized

Spoonful of Truth:

Spoon boy: "*Do not try and bend the spoon. That's impossible. Instead... only try to realize the truth*".
Neo: "*What truth*"?
Spoon boy: "*There is no spoon*".
Neo: "*There is no spoon*"?
Spoon boy: "*Then you'll see that it is not the spoon that bends, it is only yourself*". --- Spoken by the Spoon Boy character of the MATRIX [1999].

Copyright © 2013 Shawn C. Evans

All rights reserved.

ISBN-10: **1494734893**
ISBN-13: **978-1494734893**

DEDICATION

To the real magic in my life
Snug, we love you,
Always have, always will
-Dedar-
[Imagination= A shadow on a wall]

Other Books by the Author:

* **Card Through Window** – A Magical Look at the Other Side of the Glass

* **Real World Card Control Magic** – The Locators [Mnemonic Module of Moves]

* **Build Your Own Psychic Calculator and Various Mentalist Tricks**

* **Prestige Pen Projects** – DIY Gimmicks: Flash, Smoke, PSI and Other Writing Instruments

* **Psychic On Wheels** – DIY Series #3 Creative Applications

* **Mimic Money Methods** – The Magic of Bill Gaffing

* **Deceptions That Dare To Dazzle And Delight** --- The Magic of Mimesis, Volume 1

* **The Self-Contained Card Delivery System** --- Mene Tekel Magic Maximized

* **How to Hack Handcuffs Like Houdini** --- For Magicians, Law Enforcement & Urban Survivalist

Contents

Introduction

Chapter 1: Gallant Gallium

Chapter 2: Gallium Generalities

Chapter 3: Gallium Gaffs / Gimmicks

Chapter 4: Geek Gallium / LMP Alloys

Chapter 5: Cool Headed Newton

Chapter 6: Get Gallium

Chapter 7: Spoon Specifics & Etiquette

Chapter 8: Malini's Master Method

Chapter 9: The Classic Disappearing Spoon

Chapter 10: The Invisible Spoon

Chapter 11: The Penetrating Spoon

Chapter 12: The Penetrating & Restored Spoon

Chapter 13: Tuning into a Shrinking Spoon

Chapter 14: Spoon Boy Prediction

Chapter 15: Mental Melt Medallion

Chapter 16: Hitched Ring & String

Chapter 17: Coke Can Crush & Crumble

Chapter 18: Utility: Gambler's Gallium

Chapter 19: Gallium Grid Marking

Chapter 20: Goofing with Gallium

Chapter 21: Ready-Made Molds

Chapter 22: DIY Molds

Chapter 23: Portable Temp Control Devices

Appendix A: Supporting Materials

Appendix B: MSDS

Introduction

Urban Dictionary: **Geeking** – Overly excited about a simple thing: excited about something that most people would not find exciting.

Wikipedia: The word **geek** is a slang term originally used to describe odd or non-mainstream people, with different connotations ranging from an expert or enthusiast to a person heavily interested in a hobby. Its meaning has evolved to connote someone who is interested in a subject (usually intellectual or complex) for its own sake.

As defined above, *gallium* is a unique thing that most people would not find interesting. It's only the thoughtful magicians, playful pranksters and the curious minded individuals who are willing to geek-forth the secrets of this solid-to-liquid changing metal.

This book is about a special group of alloys (safe & non-toxic) that can be used to create numerous magical effects, from the vanishing of a spoon, to the melting of a coin in the hands of a spectator. Each novel application of a particular alloy is presented in detail, along with do-it-yourself projects.

For the past few years, I've been experimenting with gallium based alloys and can testify that the effects within this book are *true, tried and tested*. Each trick takes prior preparation and minimum to maximum real-time setup to effectively perform. Yet, the payoff for such groundwork is returned tenfold by way of the spectator's reactions.

I make no apologies to the class of magicians who fanatically demand that their purchased tricks are out-of-the-box, self-working gadgets with instant resets in order to be deemed practical or useful.

The preparation and application of gallium gimmicks are very practical in the framework of Max Malini's performance psychology – a mind-set employed to achieve the perception of impromptu magic at any cost in terms of time and resources.

One magician's trashed trick is another magician's coveted miracle – the same material managed with a different attitude.

Materials, Methods & Layers of Deception

In many magic effects, the main method of deception makes use of certain gimmicks or gaffs (seen or unseen), whose utility purpose is not comprehended as such from the spectator's perspective.

Throughout the process of "geeking with gallium", the materials for constructing the do-it-yourself gaffs are the actual molded alloys themselves. In other words, gallium is the gimmick and its novel application aids in creating the overall deception.

The unique nature of gallium and other low-melting-point alloys deliver the first layer of deception in creating magic effects because the majority of the public are not familiar with such exotic metals. Throughout the ages, magicians have used established principles of science combined with the newest advances in technology to astonish an uninformed audience.

Beyond this common ignorance of gallium's trickiness, the effects in this book are constructed with additional layers of deception to mystify even the informed metallurgist.

May your deceptions dare to dazzle and delight!
S. C. Evans

Disclaimer: The contents of this book are for informational and magic entertainment purposes only. The responsibility for the use of any and all information contained in this book is strictly and solely that of the user.

Chapter 1

Gallant Gallium

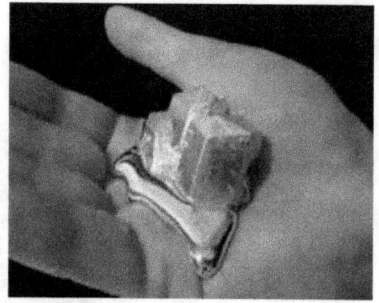

The amazing thing about the metal *Gallium* is that when it is held long enough, it will melt in the warmth of your hand at room temperature and once liquefied, it can be cooled to return to its solid form.

This distinctive transformation of gallium has practical scientific applications as well as novel uses for magicians and pranksters alike. For the purposes of this book, we'll explore the latter.

Prediction & Discovery of Gallium

In 1871, the Russian chemist Dmitri Mendeleev (co-creator of the periodic table) predicted the existence of gallium before its actual discovery. He hypothesized that the element would sit below aluminum on the periodic table and named it: *eka-aluminum*.

The chemist Paul-Emile Lecog de Boisbaudran was the first to discover gallium through a spectroscope in 1875, in Paris, France and thus the origin of the name comes from the Latin word *"Gallia"* meaning *"France"* and possibly *"gallus"* meaning *"rooster"*, which is a translation of Lecog.

The Gallic rooster was an inspirational symbol during the French Revolution and became the unofficial national symbol of France as a nation.

So, it's a good bet that you didn't imagine that you would come across a rooster while reading a book about gallium. Well… read on and you'll come across more than you anticipated.

Chapter 2

Gallium Generalities

Classification

Gallium is a chemical element categorized with the symbol **Ga** and the atomic number 31, often referred to as a post-transitional metal.

Color & Appearance

Very pure gallium has a stunning silvery-chrome color. In a solid form, gallium looks and feels metallic. When in a liquid state, the substance globules like mercury.

Beyond the hard statistics, gallium has a soothing quality, given the pleasure of watching it slowly melt-down in the warmth of your hand.

Properties

A soft, brittle metal at room temperature and a liquid metal when slightly heated. Solid gallium is soft enough to cut with a kitchen knife. Gallium is stable in air and water but will disintegrate in acids and alkalis.

Melting Point

The element **Ga** liquefies at the temperature of 29.76 °C (85.57 °F). Gallium is the only metal along with mercury, rubidium, and caesium, which can be liquid near room temperatures.

When solidifying, the metal expands and therefore should not be stored in glass or metal containers given that they may crack or break as the gallium hardens.

Misconception

Since gallium has a sharp melting-point, the common misconception is that it should have a sharp freezing point, which isn't the case.

When liquid gallium cools down below its melting / freezing point, it doesn't instantly solidify. The process gradually occurs over a range of temperatures. In other words, gallium is sharp to liquefy but slow to solidify.

In theory, gallium will stay liquid below its freezing point until "something" serves as a nucleation catalyst to start the crystallizing process. Once a crystal forms, gallium is on its way to a solid state.

Since this "something" is not fully understood by science, the best nucleation catalyst is your household cooler / freezer.

Applications

From the earliest days of its discovery, gallium was used primarily as an agent to make low-melting alloys. Today, it is mostly used in analog integrated circuits because it has semiconductor properties. Gallium arsendite (GaAs) can convert electricity into light to 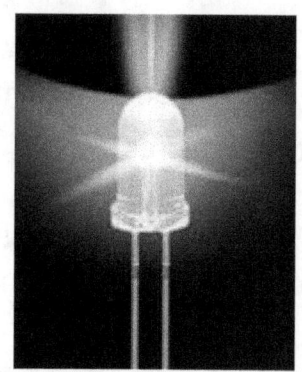 produce blue and violet LEDS (light emitting diodes) for electronic displays and digital watches.

When coated on glass, liquid gallium forms a highly reflective surface to create brilliant mirrors. High temperature thermometers utilize gallium in a liquid form.

Environmental

Pure gallium does not exist in nature as a primary mineral readily extracted from the ground but  occurs as gallium (III), found in trace amounts in bauxite and zinc ores and easily obtained by smelting. Commercially, gallium is extracted as a byproduct of zinc and aluminum production processes.

Health & Safety Issues

Gallium is considered to be non-toxic.

> For more information, see the Appendix B: Material Safety Data Sheets

Due to traces of gallium existing in the natural environment as residue on fruits and vegetables, the element can be found in the human body at very small levels --- 0.7 milligrams for an average body mass. Some commercially distributed bottled waters and vitamins contain trace amounts (less than one part per million) of gallium but with no proven health benefits for the function of the body.

The radioactive compound gallium citrate [67Ga] can be injected into the human body and used for medical scanning without harmful effects.

As indicated, gallium is not harmful in small amounts but should never be consumed in large amounts. Acute exposure to some gallium compounds such as gallium (III) chloride can cause very serious medical conditions of pulmonary edema and partial paralysis.

Gallium Staining

Gallium in a liquid form will "wet" glass and porcelain which means that it will aggressively stick to the surface of the materials. This trait has beneficial applications as discussed in chapter 18.

Gallium in a solid form will leave grayish stains on the hands when held to the point that the gallium starts to heat-up. The stains can be easily removed with soap and water.

When performing with gallium gaffs, this staining trait is not an issue when using the *immersion technique* [chapter 5] because the gaff isn't handled long enough prior to submerging it in a hot beverage.

When using the exposure technique, the gaff's melt-down is controlled and contained, as discussed in chapters 14 & 15.

Recovering Spilled Gallium

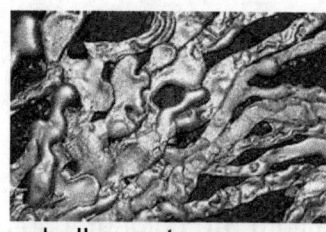

Unlike liquid mercury, separated blobs of liquid gallium will not easily re-unite on an open surface. This aspect makes any spillage of gallium a challenge to recover.

The following "spill kit" is vital in case of any accidental spillage:

* Contain the spill by covering it with an inverted cup or plastic container and wait for the liquid gallium to solidify at normal room temperatures.

When working with gallium, a good practice is to line your work area with 4 to 6 sheets of typing paper. Should some gallium spill, it will become contained on the paper.

As the spill takes on a semi-solid form, you can push the separated droplets together by taking a toothpick to roll one droplet towards another to consolidate them into a larger mass. At this point, the paper can be creased to funnel the gallium back into its holding container. The paper also helps keep the gallium clean of surface dust and dirt.

> *Tip:* Identify a droplet that has developed a crystal and use this portion to help "seed" the other droplets. If you look into a blob of liquid gallium and spot a small solid chunk, this piece is the growing crystal. Once a crystal develops, a gradual "chain reaction" causes the gallium to solidify.

* Use a syringe to "vacuum-up" the spillage.

After repeated usage, the inner walls of the syringe will become stained with gallium build-up and the tip may become clogged. This situation can be managed by removing the plunger and freeing the solidified portion. If the syringe is the lock-type, submerge it in hot water to liquefy the gallium and then pump it out into a container of cold water to instantly cool it.

* Use a "gallium mop" to swab-up the small droplets of liquid gallium. It's easier to gather / attract the liquid blobs with a solid piece of gallium.

In advance preparation, the mop is made by taking the tip of a toothpick and piecing it into a dab of liquid gallium. Once the dab has solidified, the toothpick is inserted into a cork for easy identification and handling.

Super-Cool the Gallium

* If you turn a canister of gas duster upside-down and blast spray the liquid gallium, it will instantly freeze within seconds.

Gas duster is mistakenly referred to as compressed air or canned air. The product is used for cleaning computer equipment such as keyboards and other electronic devices that cannot be cleaned with water.

For more information on Gas Dusters, see Appendix: A

Chapter 3

Gallium Gaffs / Gimmicks

"Any sufficiently advanced technology is indistinguishable from magic." --- Arthur C. Clark, *Profiles of the Future: An Inquiry into the Limits of the Possible*

For the purposes of this book, the term: "gallium gaff" means any magician's prop / object which is constructed of pure gallium or gallium based alloys as a material method to achieve a magical effect.

The Disappearing Spoon & More

A spoon made of pure gallium can be used to stage the old gag of *The Disappearing Spoon* or with an essential difference in presentation; the spoon can create the magic of vanishing, transforming, penetrating, or mind-bending.

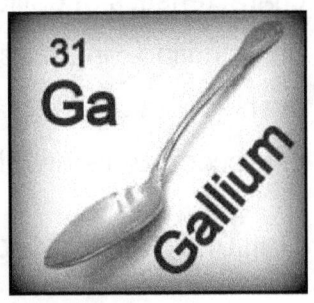

Given the properties of gallium, props made of the element are mostly limited to common objects that are perceived as "silvery metals" such as spoons and coins. You can't make gallium look like a glass bottle but you can make it appear to be a silver chalice.

A gaffed spoon made of gallium masquerades as a normal spoon because it is basically a spoon by design (though not functional). It has the same form, approximately the same weight and will sound-off with a solid clunk when placed on a table-top.

If it quacks like a spoon and walks like a spoon, it's probably a spoon.

> *Note*: In the solid form of a gaffed spoon, gallium is a brittle metal that can be handled freely during a performance. Yet, over-proving that it is a solid by forcefully dropping it onto a hard surface will result in the metal shattering into pieces.

When using gallium gaffs, the magical deception is successful because the vast majority of spectators are unfamiliar with the characteristics of gallium. Importantly, through the proper set-up, handling, and presentation of the performer (a la Malini), even spectators familiar with the concept of liquid metals may not recognize its novel use during a performance.

Gallium Based Magic

The inherent properties of these alloys dictate what type of tricks can be constructed. Basically, the types of magical effects that can be performed are:

1. Disappearances / Vanishes

The disappearance of a gallium-gaffed spoon is made possible by causing the spoon to melt within a hot or warm liquid such as coffee or tea. The liquid's holding container, usually a ceramic cup provides the suitable environment (uniform bath) for the gaffed alloy to be completely immersed in, thus causing it to immediately melt down. The opaqueness of the coffee provides the cover to hide the liquefied metal.

[See Chapters 9 &10]

2. Penetrations & Restorations

This effect is achieved by using the mechanics of #1 with a different presentation and setup. If you push the spoon through the barriers of a filled coffee cup, saucer, and table-top while immediately producing it dipping wet from the same spot under the table, then you have effected a magical penetration.

[See Chapters 11&16]

With a slightly different presentation and a little more setup, you can seemingly cause the spoon to travel through the barriers in two pieces --- first, only the bowl of the partial spoon penetrates the table-top and then the remaining handle follows in pursuit. In this demonstration, you have created the illusion of a transportation and restoration.

[See Chapter 12]

3. Transformations

This effect is achieved by using the mechanics of #1 with a different presentation and setup. If you dip a 6 inch spoon into a cup of coffee and instantly remove a 2 inch diminished spoon, then you have seemingly caused a magical transformation through alchemy.

[See Chapter 13]

4. Telekinesis

In this presentation, with the proper awareness of environment and timing, you can manage a gallium-based gaff (spoon, medallion, or coin) to actually melt or bend with suggested powers of the mind.

[See Chapters 14 &15]

Gallium and Newton's Law of Cooling

In order to use gallium in general, you have to have a basic understanding of its properties and to effectively manage the gaffs during a performance, you have to a clear understanding of the concept of *Newton's Law of Cooling*.

[See Chapter 5 for Details]

Gimmick vs. Gaff

Within the magic community, depending on whose literature you are reading, the terms: *gaff*, *gimmick*, *fake* and *feke* (older usage for *fake*) are used interchangeably.

Often for purposes of distinction, a gimmick is considered a covert device never seen, nor suspected by the audience.

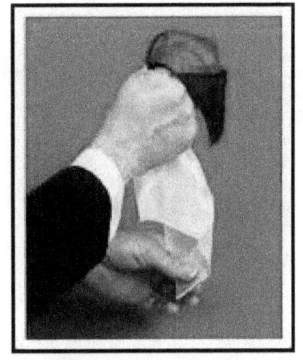

A classic example would be a dye-tube used to create a color changing silk trick. Within this book, the spoon-caddy is an example of a hidden gimmick designed to dispense a wet spoon from under a table-top.

Whereas, a gaff (fake) is a special prop camouflaged to look like a regular object -- it is seen by the spectators but not understood. An example would be a magician's thumb-tip or double-backed playing cards.

So, is a gallium spoon a fake spoon masquerading as a normal spoon or is it a real spoon gimmicked with gallium?

For the purpose of this book, any LMP alloy molded into a gallium prop (spoon, coin, ring, or medallion) will be referred to as either gaffed or gimmicked, unless noted otherwise.

Chapter 4

Geek Gallium / LMP Alloys

This chapter is a guide to a special group of LMP alloys that provide applications and options beyond pure gallium.

LMP Alloys (Non-Gallium)

Low-melting-point [LMP] alloys or "fusible" alloys are compounds that will melt at low temperatures. These types of alloys can be liquefied with a regular oven and some can even be melted in a warm environment. Most LMP metals are made of a combination of antimony, bismuth, cadmium, lead, tin, and /or indium.

LMP alloys are utilized in a variety of tool & die applications including tube bending, anchor parts, casting, and machining parts for soft metal dies.

Wood's metal is commonly used by plumbers as filler for bending thinned walled metal tubes or pipes to prevent the cylinder from collapsing. In preparation, the tubing is filled with the molten metal and after it solidifies, the tubing is bent to shape. The metal is removed by re-heating it.

> Wood's metal (aka: Lipowitz's alloy) is commercially known as Cerrobend, Bendalloy, Pewtalloy, and MCP 158 --- its melting point is 158°F (70°C).

Companies that blend LMP alloys will list their products by referring to a specific alloy's mixture and melting point, unless noted by a historical reference or brand name.

This list represents some of the non-gallium, LMP alloys:

Melting Point in F	Antimony	Bismuth	Cadmium	Lead	Tin	Indium
117F	0%	44.7%	5.3%	22.6%	8.3%	19.1%
136F	0%	49%	0%	18%	12%	21%
140F	0%	47.5%	9.5%	25.4%	12.6%	5%
140F Non-Toxic	0%	32.5%	0%	0%	16%	51%
Field's Metal- 144F	0%	32.5%	0%	0%	16.5%	51%
147F	0%	48%	9.6%	25.6%	12.8%	4%
Wood's Metal 158F	0%	50%	10%	26.7%	13.3%	0%
158-190F	0%	42.5%	8.5%	37.7%	11.3%	0%
203F	0%	52.5%	0%	32%	15.5%	0%
212F	0%	39.4%	0%	29.8%	30.8%	0%
217-440F	9%	48%	0%	28.5%	14.5%	0%
255F	0%	55.5%	0%	44.5%	0%	0%
281F	0%	58%	0%	0%	42%	0%
281-338F	0%	40%	0%	0%	60%	0%

Warning

Cadmium and lead are cumulative poisons, meaning that they can build up in your body and possibly kill you over time. Both are safe to temporarily touch and handle but the real danger comes from unwashed hands contaminating your food or inhaling airborne dust particles. The *Manufacturer Safety Data Sheets* [MSDS] and the *Toxicity Profiles* should be consulted before using any alloys mixed with cadmium or lead.

Field's Metal

A safe, non-toxic alternative to Wood's metal is Field's metal, which is a fusible alloy that becomes liquid at 144°F (62°C). It consist of indium **[In]**, bismuth **[Bi]**, and tin **[Sn]**.

Pepto-Bismol's active ingredient is bismuth subsalicylate. The famous pink formula contains bismuth metal chemically combined with salicylate.
If you're ever in need of some bismuth, you can purchase it at a gun shop in the form of shot for use in shotguns. Unlike lead shot, it is non-toxic and won't pollute the water where duck hunters gather to shoot.
Note: The latin name for tin is*: Stannum* – the reason why tin is symbolized as *Sn* and not *Tn*.

Field's metal as compared to gallium has certain pro and cons that may or may not be suited to one's preferences or performance requirements.

Cost $

For certain effects, Field's metal is a cost effective alternative for gallium. It is sold in ingot form and 4 ounces (113.39 grams) can be purchased for about $60. In comparison, pure gallium cost roughly $1.00 to $1.25 per gram.

Spoon Shine

A gaffed spoon made from Field's metal will display a polished antique silver color, whereas pure gallium will produce a spoon with a brilliant chrome look. Given that spoons come in all shades of silver, both metals are workable gaffs for casting.

Heat Source

The melting point of 144°F will require a medium heat source to liquefy the metal for casting. An electric heat gun blower is a safe (no flame), inexpensive heat source as compared to a Bunsen burner setup.

Note: A hair blow dryer is not sufficient for this task because all models are designed to stop short of reaching 140°F – the temperature at which skin burns.

Casting Procedures

When casting with Field's metal, the mold and syringe have to be heated above the metal's melting point to work effectively within a gravity-type mold. The liquid metal can prematurely solidify in the syringe before being pumped into the mold if temperatures are not maintained. On the other hand, liquid gallium is easier to cast because it is slower to solidify.

Immersion Technique

When using the old disappearing-spoon- prank or creating a trick that utilizes a hot liquid such as coffee, Field's metal is suited for the situation. The hot liquid must be served above 144°F for the alloy to liquefy when using the immersion technique (chapter 5).

Coffee Holding Time

Coffee served at 180° will have a good holding time of 11 minutes before cooling down (below 144°) to a temperature ineffective for melting Field's metal. A gaff made of gallium has a window of 30+ minutes to react to the heat of the coffee.

Durability & Handling

Compared to gallium, a prop cast of Field's metal is more durable for storage and handling. If a gallium gaff is accidentally dropped to the floor it will shatter, whereas one made of Field's metal won't.

Temperature Fluctuation Zone

Because of its near-room- temperature melting point, a gallium prop is potentially vulnerable to temperature fluctuations and must be occasionally monitored for safe-keeping. The melting point of Field's metal is outside this fluctuation zone.

Staining Issues

Field's metal does not stain. You can aggressively rub the metal and receive grey stains but this is due to normal surface oxidation.

Recap:

VARIABLES	Field's Metal	Pure Gallium
Melting Point	144 °F	85.57 °F
Cost $	0.45 /gram	$1.25 /gram
Spoon Shine	Shiny Silver	Brilliant Chrome
Heat Source	Medium	Low
Casting Procedures	Detailed	Easy
Immersion Technique	Yes	Yes
Coffee Holding Time	11 minutes	30+ Minutes
Exposure Technique	No	Yes
Temp Sensitive Zone	No	Yes
Durability & Handling	Strong	Brittle
Staining Issues	None	Yes

LMP Gallium Alloys

On the periodic table, right below gallium (atomic number 31) there is indium [element 49] and to its right is tin

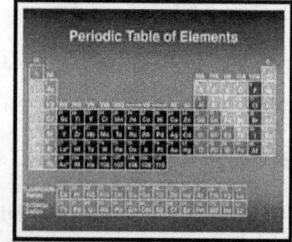

[element 50]. When these 3 elements are combined, their overall atoms bind into a compound creating lower melting-points for various combinations.

Alloys that have lower melting temperatures than pure gallium are beneficial for creating effects were the gaff employed needs a rapid melt-rate when using the exposure technique as outlined in chapter 5.

In an effect where a spoon is bent by touch or where the plot of "telekinesis" is used to melt a coin or medallion, the heat needed to cause the liquefaction is easier to manage (within a reasonable time-frame) with a gallium-indium alloy.

Here are a few gallium-alloy combinations and their respective melting points:

Alloys	Mixture Percentage % (In weight)	Melting Temp: °F /°C
#1:Pure Gallium	Ga(99.99%)	85.5 / 29.7
#2:Gallium + Indium	Ga(95%), In(5%)	77 / 25
#3:Gallium + Indium	Ga(75.5%), In(24.5%)	60 /15.7

#4: Gallium + Indium + Tin	Ga(62.5%), In (21.5%), Sn(16 %)	51 /10.7

Chilled Gaffs

As shown, alloy melting points of 77°, 60°, and 51° Fahrenheit will produce gaffs that need to be maintained in a cold-storage device (see chapter 23) prior to performing. These gaffs must be kept chilled to maintain their solid state to ensure that they do not prematurely melt-down. A familiar example is the storage of ice cubes.

Once they are exposed to environmental temperatures and heat sources, the gaffs will achieve their melting points.

Example: In summer, buildings are cooled to the range of 73°-79° (23°-26° C) and the average room temperature is about 75°F.

In theory, this environment is conducive to bending a gallium-indium [95% Ga & 5% In] spoon by touch because it will only take about 2° of generated body heat (within a few minutes) to initiate the melting point of 77° F.

> **Indoor Room Temperatures:**
>
> \> In winter buildings are warmed to the range of: 66° to 70° F (19°-21°C).
>
> \> In summer, buildings are cooled to the range of: 73°-79° F (23°-26° C).

When constructing a trick with gallium-alloys, it's important to have a clear understanding of the environmental factors that can make or break the success of a particular method.

Cut Through the Clutter

Working within the temperature sensitive ranges of gallium gaffs may seem a little confusing at first but the circumstances can be easily understood and managed by classifying them into one of two operating groups:

1: For use with the **Immersion Technique**.

2: For use with the **Exposure Technique**.

These two techniques are detailed in Chapter 5.

Temperature Teaching Tool

A digital infra-red thermometer is a good tool for monitoring all kinds of temperature surfaces when preparing, testing, or experimenting with low-melting point alloys.

These types of thermometers are easy to use -- just point the infrared light sensor towards the area or object that you want to check and click the button to read the digital measurement.

Overall, it's an ideal instrument for building an awareness of environmental temperatures as discussed throughout this book.

See Appendix: A, for an ideal pocket-sized thermometer

Here's a useful visual reference of temperature ranges:

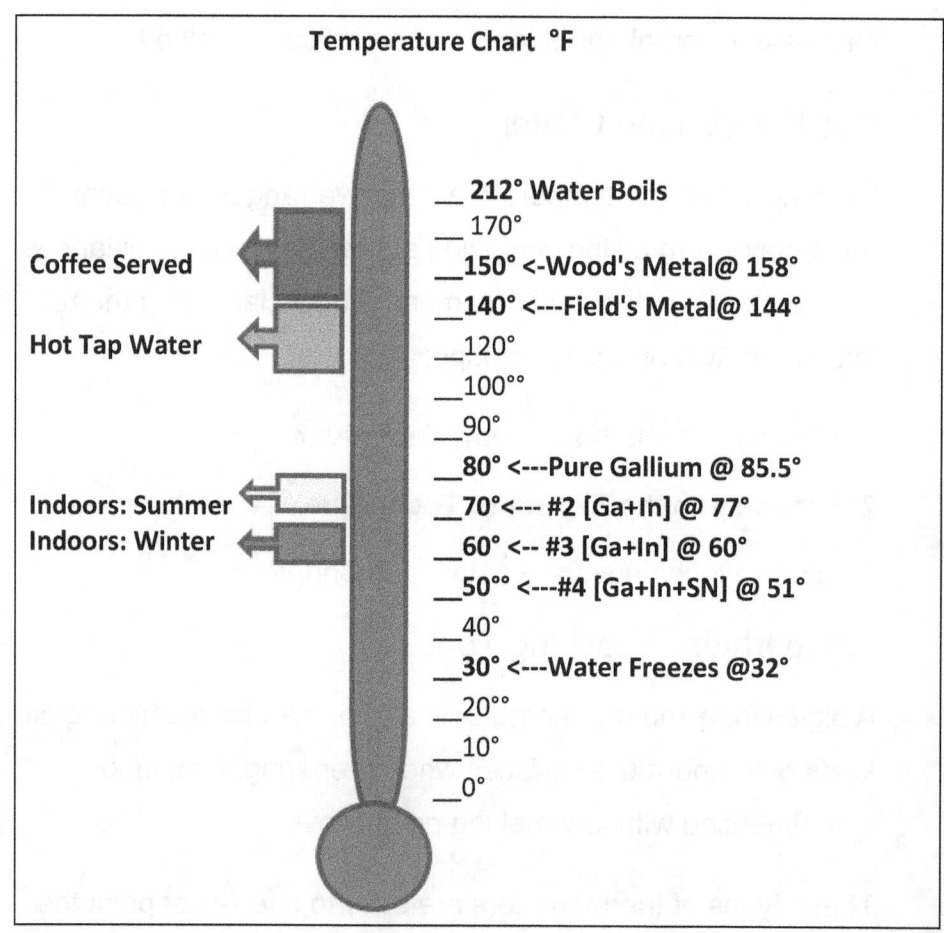

Chapter 5

It doesn't matter what temperature the room is, it's always room temperature – Steven Wright, Comedian

Cool Headed Newton

You don't have to be a mathematician, scientist, or metallurgist to use gallium but you do have to know the interaction between the material (gallium alloys) and its medium (time & temperatures) to successfully utilize it for magical creations.

The good news is that the variables of time and temperatures have been formulated by the genius of Isaac Newton as stated in his law of cooling.

Newton's Law of Cooling

The differential equation for Newton's Law of Cooling [$T(t) = T_A + (T_H - T_A) e^{-kt}$] is used to model the temperature

$$\frac{dT}{dt} = k[T - T_m]$$

$$\frac{dT}{[T - T_m]} = k \, dt$$

$$\int \frac{dT}{[T - T_m]} = k \int dt$$

$$\ln|T - T_m| = kt + c_2$$

$$e^{\ln|T - T_m|} = e^{kt+c}$$

change of an object of some temperature placed in an environment of a different temperature.

In verbal terms, the law states that the time a substance takes to cool off depends on the temperature difference between the substance and its surroundings.

In example, if you were to submerge a gaffed spoon made of Field's metal into a cup of coffee, how much time would you have to utilize the coffee before to cools down below the melting point [144°F] of the alloy, thus rendering the "coffee bath" useless?

To do the actual math, the answer depends on knowing the temperature of the coffee and the room temperature at the moment of the experiment.

Note: Most restaurants and cafes serve coffee between 170° to 200°F. According to the Specialty Coffee Association of America, coffee should be served at **180°** degrees.

So, coffee served at 180° will have a good holding time of 11 minutes before cooling down (below 144°F) to a temperature ineffective for melting Field's metal. With a pure gallium gaff, you have well over 30+ minutes.

The important concept is that an object (coffee) at a different temperature from its surroundings (room temperature) will eventually come to a common temperature with its surroundings.

In plain wording:

* A hot object cools (at a certain time rate) as it warms to its surrounding --- useful knowledge for managing the gaff's hot "bath water" such as coffee and tea.

* A cool object is warmed by its surroundings – useful knowledge for handling chilled- gaffs as listed in chapter 4.

To use gallium gaffs effectively, you must know:

A. The melting point of a particular alloy, in order to maintain it in its solid form (molded gaff) and then to transform it into a liquid state during a performance.

B. How to manage / cause the melting point to occur within a certain temperature environment.

You can manage the melting point in two ways: by rapid immersion or measured exposure:

1. Rapid Immersion Technique

In chapters 9, 10, 11, 12, & 13, & 16, the effects are accomplished by submerging the gaffed spoon or ring into a hot, uniform bath to cause it to rapidly melt down.

By far, this technique is not only the easiest to manage but it offers the best practicality when working with LMP alloys.

Variables to Anticipate:

What is the average temperature that coffee is served and at what rate will it cool off before being useless for gallium gaffs?

The molded gaff isn't under any time restraint given that it's in a "solid- friendly" climate zone (room temperature) until dipped into a filled coffee cup.

While using the immersion technique, it's the timely condition of the warm or hot liquid drink that is of primary concern for the performer. The heated beverage has to be utilized within a certain time-frame to influence the gaff.

* Field's metal and pure gallium are conducive to the immersion technique.

Good News at 11

On average, black coffee takes about 12 minutes to cool down to 143°, so when performing with Field's metal, you have about 11 minutes to utilize the coffee. If you add foam on top of the coffee, then you have 20 minutes.

It takes black coffee 30 minutes to cool down to 115°, so if you're using a pure gallium gaff, you have over a 30 minute window to use it for your performing purposes.

Within the first 15 minutes, the rate of cooling for coffee is approximately 3.08° per minute and within the last 15 minutes the rate slows to 2.16° per minute.

Metal /Alloy	Melting Point	Coffee Serving Temp	Coffee Holding Time
Field's Metal	144°F (62°C)	200°-180°F	11 Minutes
Pure Gallium	85.57 °F (29.76°C)	200°-180°F	30+ Minutes

2. Exposure Technique

In chapters 14 & 15, the effects are achieved by exposing the refrigerated gaff to room temperature and body heat.

In this method, the gaff is brought into a "hostile" environment and will be consumed because it can't "take the heat".

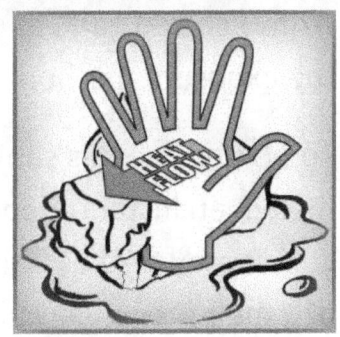

In addition to the temperature of the room, the warmth generated from body heat -- by touching and handling gaff with bare hands -- will help "massage" the melting point into action.

Average body temperature is 98.6°F as measured orally. Normal body temperature may vary 1° above or below this average. The surface area of the palm of your hand is about 90°F.

Melt-Rate: Dramatic or Boring?

Most importantly, when using the exposure technique with gallium and gallium-indium alloys, the crucial factor for constructing a magic effect is <u>the rate of melting</u>.

> If the melt-rate is slow, the process will not be dramatic enough to keep an audience engaged. Gallium will eventually melt in your hand but no amount of time control or temperature adjustments will speed up the process in a natural environment.

> If the melt-rate is rapid, the <u>progression</u> will be visual and engaging.

Pure gallium straddles the fence between the immersion and exposure techniques but the winner of the rapid-melt race goes to the gallium-indium alloys.

* Any "chilled gaff" as detailed in chapter 4, is suited to the exposure method due to its rapid melt-rate.

To be clear on this subject, a chilled gaff doesn't rapidly melt in an instant. Depending on the size and shape of the gaff, it may take as little as one minute to see indications of a melt process occurring but it will take 3-5 minutes of time-delay tactics (strategic patter and presentation) to get the overall desired effect.

These alloys have a quicker, "workable" melt process as compared to gallium for certain applications.

Chilled Gaffs

Metal /Alloy	Melting Point	Exposure Technique	Melt-Rate
Pure Gallium	85.57 °F (29.76°C)	X	Slow
#2 Ga+In	77°F (25°C)	X	Rapid
#3 Ga+In	60°F (15.7°C)	X	* Rapid*
#4 Ga+In+Sn	51°F (10.7°C)	X	Rapid

* From experimentation and observation, #3 alloy [Ga+In] is the metal best suited for molding a gaff and using it for magical effects.

> Note: It can be deduced that any gaff made-up of a Ga+In alloy could also be used within the immersion technique. However, it would be impractical to maintain such a gaff in a cold storage device and then submerge it into a hot liquid when a gaff made of Field's metal or pure gallium could be used without the additional setup.

Chapter Notes

Application Assessments

During the basic research for this book, goggling: "*How long does it take coffee to cool down*", generated a lot of weird information to compile.

Did you know that when a spoon is kept within the cup, the coffee will take longer to cool than a cup without a spoon? Well, not by much, just 0.04°. No real practical application. Nonetheless, it's a vindication for old-wives tales.

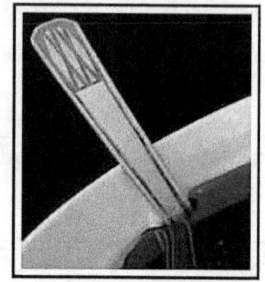

Did you know, that if you wait 310 seconds to add milk or creamer to your coffee, it will cool 85 seconds faster within a 30 minute time-frame? In the performance scheme of things, stocking creamers and cooling coffee quicker works against the immersion technique.

Which hot liquid stays the warmest within 30 minutes: coffee, tea, or water? The answer is no real difference for human perception or magical applications.

Did you know that in Newton's time, during the coffee boom in London, he once dissected a dolphin on the table of the Grecian Coffeehouse? No magical application here, unless you want to perform a routine of sawing- in- half... I won't go there.

Further Reading:

The Coffee Cooling Problem
http://mvhs.shodor.org/whatsnew/coffee/coffeecooling.html

Did Newton Drink His Coffee Black?

http://sehomeclass.bellinghamschools.org/Teachers/Henoch/AMR/newtons%20coffee%20part%201.pdf

Does Coffee Cool Faster Than Tea?

http://quicktothelab.wordpress.com/2011/03/09/does-coffee-cool-faster-than-tea/

Chapter 6

Get Gallium

There are numerous sources for individuals to purchase pure gallium in amounts from 10 grams [$17] to 500 grams [$300] and up.

* On-line sources, such as EBay.

It's usually not a good option to buy gallium from auction sites because the prices aren't that competitive and you can't ensure the purity of the metal, nor the standards for packaging it. Often, individuals resell it after having done numerous experiments and therefore it may be contaminated to some degree.

* Internet websites:

Currently, there are two good sources that cater to individuals for small or large quantities of gallium:

www.rotometals.com

www.galliumsource.com

Here's a list of some of the stock available:

Metal / Alloy	Mixture % (In weight)	Unit Sold	Cost $
#1: Pure Gallium	Ga (99.99%)	Grams	On average: $1.25 per gram
#2: Gallium + Indium	Ga (95%), In (5%)	Grams	Custom
#3: Gallium + Indium	Ga (75.5%), In (24.5%)	Grams	Custom
#4: Gallium + Indium + Tin	Ga (62.5%), In (21.5%), Sn (16 %)	Grams	Custom
Field's Metal	In (51%), Bi (32.5%), Sn (16.5%)	Ingot-4 ounces (113.39 grams)	$60

Alloys #2, #3 & #4 are currently custom-made alloys that can be purchased via request by contacting the *Gallium Source* on-line.

> *Note:* A very common ratio for this material is: [Ga 68.5%, In 21.5%, & Sn 10%], called *Galinstan*, a registered trade name. Galinstan is not useful for molding a gaff given that it has a melting point of −19 °C (−2 °F).

The website *DisappearingSpoons.com* sells one ready-made alloy [gallium+ indium] with the melting point of 58°F.

Another option for the DIY hobbyist is to mix your own materials. Gallium, indium, and tin are bought in the proper quantities and the ratios for #2, #3, or #4 are mixed by weight and smelted.

Buying Gallium & LMP Alloys

When purchasing gallium, you should look for the 3-P's:

* *Purity of the Metal*: 99.9%.

* *Packaging of the Product*:

Professional suppliers of gallium know its properties and therefore will ship it safely. The metal is usually prepared and poured into a leak-proof Kautex plastic bottle, then bagged with absorbents. Gallium expands 3.1% when it solidifies and that's why small amounts are only partially filled in bottles to ensure that the expansion doesn't cause a break in the holding container.

The marketplace has made the overall pricing of gallium competitive, yet the shipping costs vary to a large degree and can be a source for cost savings. This situation is due to different companies interpreting how the metal should be shipped according to Hazmat Standards.

* *PayPal Type Protection:*

This isn't an endorsement for any company per se, just a buyer's caveat should anything go wrong with your on-line order; you should have options for a resolution or a refund.

Chapter Notes

When you receive your package through the mail, the gallium most likely will be in a liquid state due to having traveled through various time and temperature zones. Upon opening the bottle, you'll notice that the inner walls of the container will be mostly stained with a thin lining of gallium as a result of the topsy-turvy handling of the package.

These stains can be recovered by placing the bottled in a refrigerator for about 15 minutes until the main bulk of gallium is solidified. At this point, the sides of the bottle can be tapped and squeezed to cause the stains to flake off the walls. It's a small amount in terms of weight but it's worth the effort.

Chapter 7

Spoon Specifics & Etiquette

They dined on mince, and slices of quince, which they ate with a runcible spoon; And hand in hand, on the edge of the sand, They danced by the light of the moon. --- Edward Lear, The Owl and the Pussy-Cat

This chapter examines the world of spoons in order to identify a generic spoon to be used as the master for your spoon mold.

If you're going to make a gallium spoon to prank someone with the old disappearing-spoon-gag, then any spoon will do.

For magical presentations you want to select a spoon style that will masquerade as a common spoon found in homes or at restaurants because a near-matching gaffed spoon can be switched for a borrowed spoon to "sell" the overall deception of a magical event done with a common item.

As a perfectionist, you may choose to make one or more molds for your choice of pre-identified spoons. Or as a walk-around performer you may choose to identify a generic spoon that can be employed in a number of impromptu settings. Such a spoon is described within this chapter.

Spoons to Sip, Sap, & Sup

A spoon's style is usually named after the food or drink for which it is most often used, the material from which it is made or a main feature of its structure.

Google "spoon types" and you'll get a long list of varieties: bouillon, caviar, Chinese, coffee, cutty, demitasse, dessert, egg, grapefruit, horn, iced tea, marrow, melon, parfait, runcible, salt, saucier, soup, teaspoon, tablespoon, etc.

Spoons found in restaurant and cafe settings:

* Place spoon --- an all-purpose spoon with an oval bowl. Length range: 6.5 to 7.5 inches. The most common found is 7 inches [183mm] in length.

* Iced-beverage / ice-tea spoon --- used to stir cold beverages served in a tall glass. Length range: 7 to 10 inches.

* Café coffee spoons --- these spoons range in length from 3.5 to 5 inches.

Common spoon used in home dinning:

* Teaspoon --- used to stir hot beverages, sip soup and eat solid food. The average teaspoon measures approximately

5½ to 6¼ inches in length. The most common found is 6.25 inches [158mm] in length.

Conclusion:

A generic spoon would be one with a length of 6.25 inches (teaspoon) or 6.5 inches (place spoon) with a plain, utility pattern.

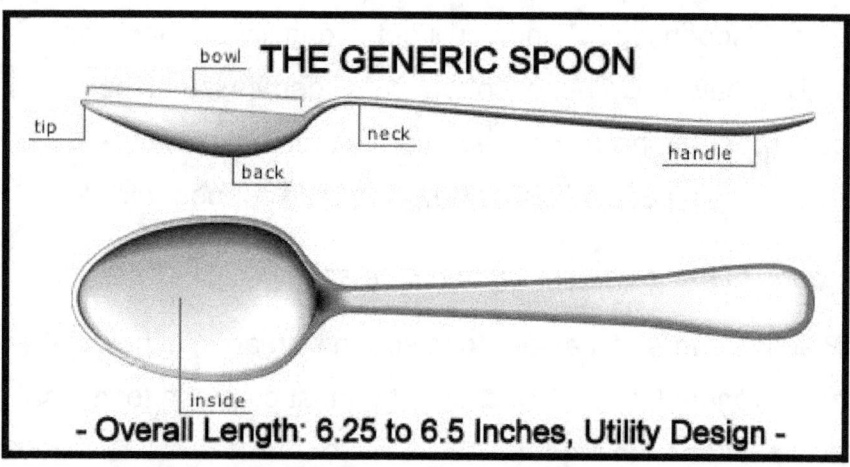

- Overall Length: 6.25 to 6.5 Inches, Utility Design -

Note: Throughout this book, the various parts of the spoon [tip, bowl, back, neck & handle] are identified for purposes of distinction.

A mold formed with a plain teaspoon or a place-spoon can be used to create a generic, gallium- gaffed spoon --- one that would pass for those found in homes, diners, cafeterias and restaurants.

Amount of Gallium Needed for a Generic Mold

How many grams of gallium are needed to fill a spoon-mold? The simple answer is however much gallium it takes to fill the cavity of the mold.

Once you cast your mold with the spoon that you have selected as your generic model, you are basically stuck with that configuration in grams.

Prior Consideration & Calculation

Not all similar looking spoons have the same weight. In example, a quality spoon (6" long & <u>thickly</u> made) weighing 1.15 ounces, will take about 33 grams of gallium for molding a model. In comparison, a cheap spoon of the same shape and length (6" long & <u>thinly</u> made), weighting in at 0.55 ounces will take about 16 grams of gallium, plus 4 extra grams for good measure.

> *Note*: The bulk of the weight difference between a quality spoon and a cheaper version is the bowl depth. A quality spoon has a deeper bowl than does the cheaper one, which has a smaller recess for the basin.

Given that gallium is costly to purchase, a good factor in choosing a generic spoon is to weigh it in grams to calculate how much gallium will be needed before casting your actual mold.

If your ideal spoon is "gram heavy", you can use a *DREMEL* to thin out the thicker length of area between the neck and handle of the spoon. You can also thin out the sides of the handle.

Of course, a gram of stainless steel in a typical spoon is not identical in atomic weight and mass to a gram of gallium. The above gram-to-gram method of calculating the amount of gallium needed prior to casting the mold is a very "rough & ready" approach.

In general:

When casting a spoon mold, a typical size spoon will create a cavity that will take 20-25 grams of liquid gallium to fill.

Various suppliers of ready-made spoon molds (chapter 21) will suggest a range between 16-20 grams of gallium. This amount is calculated on their specific mold designed with their choice of spoon -- often the thinnest available to control the cost of gallium needed.

To eliminate costly trail & error attempts at purchasing a sufficient amount of gallium prior to building an actual mold, a minimum of 20 grams and a maximum of 30 grams are recommended.

The ideal purchase is 25 grams when the amount of gallium can be purchase in multiples of five grams.

Chapter 8

Malini's Master Method

Terms: **Impromptu Magic**: Tricks that can be performed *promptly,* without lengthy setups, elaborate props or gimmicks. Magic that takes advantage of common objects that are available at the moment.

This chapter is situated between the chapters on gallium and the chapters on magic effects because Malini is the glue that holds this book together.

The Man & Legend

Max Malini (August 14, 1873 – October 3, 1942) is acknowledged as one of the all-time masters of impromptu magic.

Tales of his "spontaneous exploits" in bars, restaurants, and before members of Congress have become the stuff of legend.

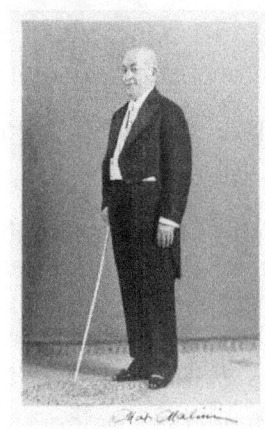

Even though he had small hands, Malini perfected his sleight-of-hand techniques with timing, psychology, misdirection and advance preparation to become known around the world as a performer who could seemingly create unrehearsed miracles out of nowhere.

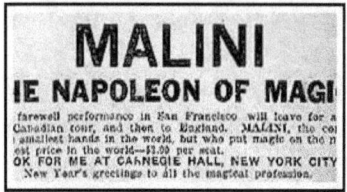

Technically speaking, Malini's most famous feats weren't actually impromptu by definition. However, his effects were brilliantly executed to create the perception of being unplanned.

Prior Planning Promotes Phenomenal Performances

Malini's genius was his attitude and application of setting-up a trick days, weeks, or months in advance while patiently waiting for the right moment to undertake the performance.

In one of his renowned effects, Malini would challenge the spectators to a guessing game. Then he would borrow a hat; use it to cover a flipped coin, while asking, *"Heads or tails?"* When the hat was lifted, a block of ice would be sitting on the table.

To this day, the magic community doesn't know exactly how Malini concealed and loaded the ice. What is known is that such a feat takes advance preparation and timing to present as a spur-of-the-moment event.

Thus, one of the real secrets of magic, which is a trick-within-a-trick, is the method of using prior planning to create the illusion that a trick is impromptu.

This technique of "contrived spontaneity" is the difference between a good trick and a full blown miracle. When spectators are convinced that a trick was done in the moment and without gimmicks, they are willing to believe that something remarkable has occurred.

A gallium spoon can be used to simply prank someone or with some planning, patience (a la Malini) and presentational skills, it can be employed to create miracle status effects.

The Mechanics & Dynamics of a Miracle

The utility of a gallium gaff is maximized by:

1. Prior Planning & Set-up Phase.

When asked about his magic, Malini responded to fellow magicians that the real secret to his success wasn't his technical skills or bold personality but rather *"rehearsal".*

Like any competent magician, Malini practiced the various sleights and moves needed to execute the *mechanics* of a given trick.

Unlike many run-of-the-mill magicians, his "rehearsal technique" was in planning for the *dynamics* of the trick -- the aspects of where to set it up, when and how to present it in a manner to create the illusion of spontaneity.

*This mindset can be cultivated by using the **5-T's**:*

Task: The goal is to regularly re-evaluate and revitalized your magic to appear natural and impromptu.

Technique: Any system (sleights, moves, misdirection, patter) that lends itself to the task.

Tools: The main tool is applying thoughtfulness to the magician's toolkit. All apparatus [props, gaffs, gimmicks, gizmos] must consist of or represent common objects.

Timing: Within the art of magic, timing has many applications from time-delays to the seamless execution of sleights. In the spirit of Malini, timing is the proper introduction of a trick to give it the appearance of spontaneity.

Tracking: This is the use of feedback to evaluate to what degree your task has succeeded. When performing, the mechanics of a trick should be completely on auto-pilot so as to allow the performer to evaluate the spectator's reactions.

While the participants are watching the magician, the magician should be watching the audience to measure the impact of the trick throughout the progress of the routine. Was the overall response indifferent, good, or spectacular? Did the spectators become suspicious at any point in the presentation? Should you make adjustments?

2. Utilizing Common Objects (Gaffed or Un-gaffed)

From the start, a gallium gaff is made to pass as a common object such as a spoon. The thoughtful magician may go so far as to custom-mold a spoon to match the design of an antique spoon or the types served at ritzy restaurants.

This added effort is to "sell the illusion" that the gaffed spoon used during the performance is real and borrowed.

However clever or perfectly crafted a gaffed prop may be, it cannot stand alone as the sole method to finalize the deception. The gaff needs to be cloaked within the proper handling and presentation to be truly invisible to human senses.

In example:

When a magician borrows an object, the implied message is that it's not gimmicked or "tricked-out" in any manner as to assist the performer.

When an object such as a planted spoon is "borrowed" from its natural habitat (dining table, coffee shop, restaurant silverware), the psychology is the same but more importantly, the setup never leads to over-proving.

From the start, the use of common objects (gaffed or un-gaffed) as props, will aid in the perception that the magic is spur-of-the-moment.

Malini traveled around the world entertaining royalty and heads of state with simple "pocket props" such as cards, coins, and personal items. Had he known about gallium, he surely would have gaffed some spoons.

3. Pursuit and Timing Phase.

During this phase, the gaff is loaded. It is planted in position, strategically placed on the table or switched for someone else's dinnerware. The spoon "lies in wait" for the proper moment to bring it into play – a moment that will "sell the illusion" that the magic is unplanned.

How long do you wait for the right opportunity? To paraphrase Malini in his native accent: "*I vait, I vait a veek*".

Malini's notorious patience for presenting a trick at the right moment wasn't a passive process of waiting but rather the active pursuit of seizing the right opportunity.

Here are some tips to seize the right moment:

> Wait until you're asked. If you're well-known as a magician, invariably someone will ask you to perform. Upon such a request, hesitate and act as if you are totally unprepared to do so. A magician eager to perform smells like a setup.

> The proper moment would be one in which the dinner conversation naturally turns to a subject that the performer can seize upon as an introduction to perform. Active listening is a magician's tool.

> To steer the conversation, the performer plants keywords or phrases to subliminally direct the conversation. You don't have to be a professional psycho-hypnotist to understand the power of words to guide suggestions, clues, and hints.

> To steer the conversation with a visual clue, a regular spoon is slightly but noticeably bent and placed next to a talkative guest. When the spoon is discovered, it's a sure thing that the subject will turn to bent spoons and psychic powers.

> One method of creating the perception that a trick is impromptu is to perform it as an afterthought. When performing a few tricks, you act as if you're finished and then you remark: *"That reminds me of something special"*.

> Another method is to use a stooge to publically challenge you to do a trick. The stooge talks about how he heard that you could do *"so and so"* and then he dares you to do it.

Overall, this main phase is conducted to produce convincers -- added layers of deception-- to eliminate any suspicion that a gimmick is being employed and to generate the perception that nothing has been setup or planned.

As Malini understood, the real secret to successful impromptu magic is paradoxically prior planning.

Chapter 9

The Classic Disappearing Spoon

Effect: As a colleague uses a spoon to stir his coffee, it instantly dissolves, disappearing into the cup. The unsuspecting victim is left holding just the end tip of the spoon's handle.

Historically, this practical joke has come to be known as *The Disappearing Spoon Trick,* a prank traditionally played within the laboratories of chemists and metallurgists.

For a brief time during the 1950's, a version of this prank was made available to the public until it was discovered that the marketed spoon was made of toxic material.

An earlier explanation of this gag was recorded in *The Playbook of Metals* [Chapter XV, Experiments with Bismuth, page 439] by John Henry Pepper (1861):

The fusible metal that melts when placed in boiling water consists of eight bismuth, five lead, and three tin, and some persons—of course wags—have even gone to the expense of having it made into spoons; and the surprise of any grave personage who uses such a base spoon to stir his or her tea, to see that usually solid article of plate melt away before their eyes, may be more readily conceived than described.

Pepper's book includes narratives of visits to tin, lead, copper, and coal mines; along with numerous experiments relating to alchemy and the chemistry of the fifty metallic elements for that time.

For the magic historian, Pepper is mostly remembered for

developing a technique known as *Pepper's Ghost,* a method

of projecting an actor onto a stage using a sheet of glass and an ingenious use of lighting. The projected actor had an ethereal, ghost-like appearance and was perceived to be performing alongside the other actors.

Demonstrations of the "ghost effect" were so popular that they produced a steady stream of business for the theatres as intrigued people returned to the shows in an attempt to figure out the method being used. At the time, the famed physicist Michael Faraday was so amazed and puzzled that he requested an explanation.

Recently, my first introduction to the properties of gallium came from Sam Kean's wonderful book, *The Disappearing Spoon: And Other True Tales of Madness, Love, and the History of the World from the Periodic Table of the Elements (2010).*

The book's actual advertisement reads:

The Periodic Table is a crowning scientific achievement, but it's also a treasure trove of adventure, betrayal, and obsession. These fascinating tales follow every element on the table as they play out their parts in human history, and in the lives of the (frequently) mad scientists who discovered them. THE DISAPPEARING SPOON masterfully fuses science with the classic lore of invention, investigation, and discovery--from the Big Bang through the end of time.

Well, we've gone from spoons to spooks to Sam's stuff within a short space. So now, let's take a look at pranking *The Disappearing Spoon*.

The Mechanics of the Prank

* **Materials Needed:** Gaffed spoon and an available cup of hot or warm beverage.

> Option 1, Metal: Molded spoon made of Field's metal.

> Option 2, Metal: Molded spoon made of pure gallium.

* **Setup**: Minimum, on-the-spot arrangement. The gaffed spoon is planted in place or switched for a regular spoon.

* **Melting-Point Method**: Immersion Technique [Chapter 5].

* **Time &Temperature Considerations:**

Metal /Alloy	Melting Point	Coffee Serving Temp	Coffee Holding Time
Field's Metal	144°F (62°C)	200°-180°F	11 Minutes
Pure Gallium	85.57 °F (29.76°C)	200°-180°F	30 Minutes

* **Presentation:** This effect is a prank and depends on the potential victim using the gaffed spoon while stirring it in hot coffee or tea.

To ensure that this happens, the prankster patters about how the molecules in certain metals become softer when stirred (within a liquid) in a counter-clockwise direction as compared to a clockwise direction. The victim is requested to do so with his available spoon.

Another story-line is to ask those present to help conduct an experiment to determine whether coffee will stay warmer with or without the spoon in the cup. As the participant uses his spoon to test his theory, it immediately dissolves within the cup.

* **Caution**: Even though the gaffed spoon is made of a non-toxic alloy, the best consideration is to have the spoon stirred in your cup of beverage. There's no reason to lose a friendship over messing-up someone's morning addiction.

* **Recovery of Liquid Metal:** Overt. Since the effect is a prank, you can openly recover the liquefied metal at your convenience.

Chapter 10

The Invisible Spoon

The following 15 minute routine uses everyday objects at the dinner table for a very visual presentation at each interactive phase. Each of the 6 modules flow into the next phase, yet the total performance can be shortened per the performer's preference or time constraints.

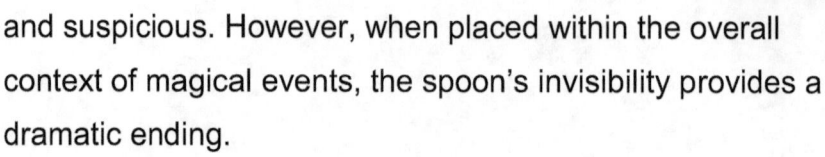

At least 3 of the modules should be performed before executing the "invisibility" of the spoon [module 6]. Without the supporting modules, the spoon's disappearance becomes problematic and suspicious. However, when placed within the overall context of magical events, the spoon's invisibility provides a dramatic ending.

Main Effect

Once coffee is served at the dinner table or in a restaurant setting, the conversation casually turns to *ESP* and *Psychic Phenomenon*. Being the Magician, you are asked whether or not you can bend spoons? You remark that such a feat is interesting but more interesting is the ability to affect matter at the molecular level.

You demonstrate such by:

* Creating gunpowder sparks.
* Instantly solidifying a glass of water.
* Instantly making snow.
* Turning water into wine.
* Making a coin travel through space and time.
* Robbing a spoon of its visibility.

Materials Needed

During the entire routine, you'll need 2 prepared packets of sugar (loaded with Slush Powder), 2 prepared packets of *Sweet & Low* (loaded with Snow Powder), 1 prepared packet of sugar (loaded with a dime), preparations for producing "wine"  and one gallium gaffed spoon. Each item will be described in detail per module.

Presentation & Performance

One of the most important things to know about close-up magic done at the dinner table (in a home or restaurant setting) is that during the time spent conversing and eating, you have all the opportunity to setup your tricks.

Setting-up a routine of effects during meal time is possible because real- world misdirection is a natural part of the performing conditions. Looking at the menu, passing the salt, handling the silverware, food being served and waiters approaching the table are ideal situations that facilitate opportunities for loading, planting, concealing and lapping.

Module1: The Sparkle Effect

* Method: Most people are unaware that *Sweet & Low*, when sprinkled over a flame will sparkle similar to gunpowder --- not as pronounced but it will certainly spark when consumed by the flame. This is a natural property of the sweetener. This effect is done first because you can freely borrow any packet, which helps generate the impression that everything has been borrowed throughout the routine.

* Action Step: Have anyone take a packet of *Sweet & Low*, tear the corner open and lightly sprinkle it over a flame. You provide the flame via candle, matches or a pocket lighter. As they sprinkle the contents of the packet, make sure that you

snap your fingers or make a magical gesture that causes the sparks to happen.

> *Note*: If you can produce the flame in a magical fashion by way of a match-pull, this will enhance the overall trick. Small enhancements to an effect are often the difference between doing a trick vs. producing a miracle in the minds of the spectators.

* Patter: "*Ok, let me demonstrate how to affect matter at the molecular level. We'll take a packet of Sweet & Low, make a flame, snap my fingers and ... we have caused Sweet & Low to sparkle. Alchemist of old would be de-lighted.*" [Pun intended].

Immediately proceed with the next effect:

Module 2: The Jell-O Effect:

* Method: This effect is done by using *Slush Powder* which is a chemical that will instantly turn a glass of water into a solid mass of gelatin. The powder is used to create numerous effects in magic and the material is mostly hidden from view. In this routine it is openly used to create a highly visual transformation.

* Setup: Prior to starting this effect, you covertly plant two prepared sugar packets into the condiment holder.

You have the prepared packets thumb clipped in your hand and as you reach to get some sugar, you leave behind the loaded packets.

> *Note:* A thumb clip is where you have an item in the palm of your hand and you use your thumb to hold (clip) the item against the palm so that when the hand is held palm-down with the fingers outstretched, the item is hidden from view.

Of course, the sugar packets have been prepared well in advance by emptying out the sugar and replacing the contents with *Slush Powder* or *Snow Powder*.

> See end of chapter for how to prepare packets of *Slush* & *Snow Powder*

All of the prepared *slush & snow powder* packets can be covertly stacked into a condiment holder at the start of the routine. It's expedient to fold or cease the corners of all the preloaded packets so that they can be identified when needed and tracked to ensure that no one accidentally uses them.

* Action Step: You have a spectator hold a glass of water while you casually pick up the two gaffed packets of sugar from the condiment holder. You have the spectator take a

knife or straw and stir the water in a <u>clockwise</u> direction. As you pour the sugar packets into the water, it instantly transforms into a solid mass of gel.

> The whole business about stirring the water in a clockwise fashion is used as a magical-gesture-technique which conveys that you have a secret knowledge that allows you to perform the magical transformation. Having the spectator participate by holding the glass adds to the interaction of the effect.

* Patter: *"We can affect matter even more, by taking a glass of water and ... let's see, some sugar packets. We'll stir the water in a clockwise direction. The secret method of transformation is to stir in a clockwise direction. And... with a magical wave, we have transformed the molecules of water into... Jell-O."*

Proceed with the following effect:

Module 3: The Snow Effect:

* Method: This effect is done by using *Insta-Snow,* which is a chemical that will instantly turn a glass of water into what appears to be snow.

* Set-up: You'll need to prepare in advance two packets of *Sweet & Low* (or a similar sugar substitute) with *Snow Powder.* Most restaurants have a variety of sugar and sugar substitute brands which is favorable for advance preparation and being able to track and differentiate between packets.

> Another option for this phase of the routine is to use a wrapped drinking straw to dispense the *Snow Powder* --- see end of chapter for the details.

* Action Step: As the glass of *Jell-O* is finally passed around for viewing, you have another spectator hold a glass of water. As the *Sweet & Low* packets are emptied, you have the spectator take a knife or straw to stir the water in a <u>counter-clockwise</u> direction. As the spectator stirs the water, it instantly turns to *snow*.

> *Note:* It's important to wait for everyone to view and inspect the glass of *Jell-O* before doing the *Snow Effect*. If you don't, some spectator's will miss the start of the current effect. The proper timing will allow everyone to comprehend what's going on. Many good effects have been ruined by assuming that spectators are paying attention at the right time or they comprehend what's going on. Spectator management is simply directing attention at the right time and in the right place without making assumptions as to what they are seeing or not seeing.

* Patter: *"Ok, so far, we've affected matter at the molecular level with transformations of sparkle and Jell-O. Let's see what will happen if we just stir water in a counter-clockwise direction with a magical wave of the hand and an extra snap of the fingers... snow"!*

Module 4: The Wine Effect:

This is a good time to wait (a la Malini) in order to prompt further conversations as to what has happened. Invariably, after the *snow* has been produced, someone will say something to the effect: *"Well, that's great but can you turn wine into water?"* And then you do just that. The spectator's reactions are truly magical.

* Action Step: You take an "empty" glass and pour water into it. As the water fills the glass, it instantly turns to wine.

* Patter: *"Oh, turning water into wine is an easy one. I get request for it all the time. You just pour water into an empty glass and give three magical snaps of the fingers and... instant wine!"*

* Method: This effect is done by using sugar-free *Kool-Aid*. In the past, I used a clothing dye (*Rit-dye*, Purple 13) to simulate wine, but the liquid couldn't be consumed because of the chemicals involved. By using *Kool-Aid,* you can take sip of the wine and comment on the great taste to add to the realism of the effect. Another option is to use *Easter Egg Die Tablets*.

* Setup. You setup this effect at the very beginning of the entire routine before you do THE SPARKLE EFFECT. After you drink a glass of water (leaving just enough water to cover the bottom surface of the glass), you covertly drop the

die tablets onto the bottom of your glass. Within minutes the tablets will dissolve. The tablets have to be of a wine color and the type that doesn't need vinegar to dissolve.

If using *Kool-Aid*, you can covertly dispense the powder by using a magician's *Thumb-Tip*. For pocket management, a cork is used to plug the contents of the thumb- tip before using it.

If you're concerned about this setup being spotted, you can position the glass behind tableware (condiment holders, shakers, napkin dispensers). The glass "sits" ready to be used. You have to be alert to anyone trying to clear the table of empty glasses.

> *Notes on Managing Water Glasses:* As this point in the routine, you will have needed two glasses of water and one empty glass. The setup of having at least 3 filled glasses available can be anticipated by noting how many guests are at the table. If you have 3 or more guest, the glasses are ensured to be available. Most people don't drink all of their water served at the beginning of the meal because they have another beverage coming. Also, you can request one or two glasses of water for "non-existent" guests --- who may or may not be arriving.

Module 5: Space & Time Effect (Dime to Sugar Packet)

At this point you have magically produced sparkle, *Jell-O*, snow and wine. Now it's time to bend space using the classic *Dime to Sugar Packet* effect.

> See end of chapter for how to prepare packets of *Sealed Sugar with Dime*

* Action Step: Casually grab a sugar packet from the condiment holder and ask a spectator to safely hold on to it with a closed fist.

Next, you ask to borrow any dime. Display the dime in your open palm, read off the date and ask another spectator to help you by making a mental note of it.

You vanish the dime and have the spectator immediately open the packet of held sugar. When the contents of the packet are emptied, a dime drops out with the sugar. The date matches!

* Method: The sealed sugar packet (with dime) has been pre-loaded into the condiment holder.

> See end of chapter for how to prepare packets of *Sealed Sugar with Dime*

You borrow a dime and misread the date. The date may actually be *1995* but you read out-loud the date of *1990* (the date of the dime in your prepared sugar packet). Then, you do a Slydini type vanish or your favorite type of coin vanish with the borrowed dime.

> See end of chapter for *How to Do a Slydini Coin Vanish*

* Patter: *"Well, I didn't bend any spoons, but maybe I can demonstrate how to bend space and time. Will someone hold onto this sugar packet? Can I also borrow a dime from anyone? Thanks, 1990, a good year. I think that's the date our glass of wine was made? Please hold your fist tightly. Like this. Like the way I'm holding onto my dime. So tight, that it has… dematerialized…VANISHED. Please open up your packet of sugar and slowly pour out the contents. There's a dime? What's the date on the dime? 1990! See, time can fly. Space and time have been bent!"*

Module 6: The Invisible Spoon

This last module is the climax of the routine. This is the point in the routine that you "rob the spoon of its visibility". You do not cause it to vanish.

After finishing module-5, as an afterthought, you borrow a spoon and try to bend it with your mind as requested earlier. When failing to do so, you notice the richly black color of your cup of coffee. That it's the darkest color which is the result of the absence of the absorption of light.

Seizing upon a moment of inspiration, you stir the spoon <u>clockwise & counter-clockwise</u> into the cup, in order to agitate the molecular structure as in the previous demonstrations.

As you remove your hand, the spoon has become invisible because the coffee's color has stolen the spoon's visual wavelength, rendering it undetectable to the human eye. You take the invisible spoon and finally bend it for all to imagine.

* Melting-Point Method: Immersion Technique [Chapter 5]. The spoon is made of gallium and is planted in place at the start of the routine. The main routine should begin once the coffee is served. From that moment, you have 30 minutes to utilize the coffee's heat to melt your gallium gaff. The total routine, modules 1-6, takes less than 15 minutes to execute.

* Patter: *"As you can see, we've changed the molecular structure of quite a few things. Maybe I can bend a spoon as requested earlier. Now that I notice the reflection of the spoon in the coffee, maybe I can do two things with it. I'll use the warmth of the coffee to soften the molecules and then I'll use the blackness to rob the spoon of its visibility. Since the color black is the absence of the absorption of light, it often attempts to steal light from other objects by messing with its wavelength. Look, the spoon has been robbed of it visibility! Here... to finish my task, I'll finally bend it for all to see".*

How to Prepare Packets of Slush & Snow Powder

> Take a packet of sugar or sugar substitute and shake the contents so that that the bulk of the sugar is crammed to the bottom end of the packet. This lump will allow you to cut open the packet without slicing through to the other side.

> Cut a vertical line across the lump, staying within the frame of the packet or the logo design in order to camouflage the cut.

> Pour out the contents.

> Replace with *Slush* or *Snow Powder*.

> Reseal the packet with *Elmer's School Glue*.

> To identify the packet, fold and crease one or two of the corners.

When making packets of both powders, be sure to remember which brand has which powder.

Note: It takes about 1 teaspoon (1 heaping packet) of *Slush* or *Snow Powder* per 8 ounces of water. As a rule, if the glass of water is wide & tall and about full, you'll need two packets of the same powder. If the glass of water is normal in size (holding about 8 ounces of water) then a packet will do. When in doubt, use two packets.

How to Prepare Packet for Sealed Sugar with Dime

> Take a packet of sugar and shake the contents so that the bulk of the sugar is crammed to the bottom side of the packet.

This lump will allow you to cut open the packet without slicing through to the other side.

> Cut a vertical line across the lump, staying within the frame of the packet or the logo design in order to camouflage the cut.

> Slide a dime (make note of the date) into the opening and reseal with glue. You don't remove the sugar from the packet.

> To identify the packet, uniquely fold one or more of the corners.

How to Dispense Powders with Wrapped Straws

Besides using loaded sugar packets to deliver *Slush & Snow Powder*, for diversity of method you can use wrapped drinking straws.

> Take a wrapped drinking straw and make a small slit at the top end (we'll call this end: A) of the straw. The slit is made at the side of the wrapper and only large enough to allow the plastic straw to be removed. Save the wrapper.

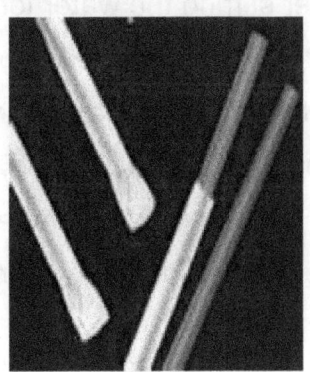

Note: The straw should be of a solid, non-transparent black or red color.

> Take a wad of paper and plug the bottom end of the straw (end: B). The wad should be pushed up into the straw about half an inch. This plug will keep the powder from spilling out of the bottom.

> Fill *Slush or Snow* powder into the top end (A) of the straw. The powder should fill ¾ the length of the straw.

> Place the straw back into the wrapper and seal the slit with glue. Make a small identifying mark on the wrapper to indicate the proper orientation of the straw with end: A, upward. This will prevent the powder from spilling out of the top end of the straw until needed.

During the performance, you take the straw (A-end upward) and remove the wrapper. Next, stir the water with the end (B) of the straw. Do not submerge the straw beyond ½ an inch.

After stirring the water for a moment, remove the straw and turn it over end-to-end and submerge the top end (A) into the water and blow through the bottom end (B) of the straw. Even though end-B is plugged, enough air can be blown through the straw to dispense the powder into the water.

Do not blow excessively hard because you will create an air bubble that will splatter the powder and water. As the powder gels in the glass, leave the straw standing in the middle of the solidified water for a magical display.

How to do a Slydini Vanish

> With your left hand, place a dime in the center of your right palm.

> While placing the dime, position you right- hand about two inches above the edge of the table with your finger tips about two inches overlapping the edge of the table.

At this point, you're going to turn your right- hand palm down, letting the dime drop into your lap. As you make a fist with the same continuous motion, you extend your hand towards the spectator who is holding the sugar packet.

When you make a fist (after turning your hand palm down), it appears as if you are merely grabbing hold of the dime (which gravity has landed in your lap).

> The straightforward movement of your hand and arm naturally covers the dime being lapped. The larger motion (extending your hand) covers the smaller motion (the dime being lapped).

For added misdirection, look at the dime. Then look up and make eye contact with the spectator. As you execute the move, say: *"I want you to hold the sugar packet as tight as I'm holding this dime"*.

The move that you just made is justified in that you are showing the spectator how to hold the sugar packet by example.

Chemical Information:

> Insta-Snow is NOT a food grade product. DO NOT INGEST or INHALE. Keep out of the reach of children 9 & under. Material is non-toxic. If powder gets in eyes, flush with water.

> Slush Powder is NOT a food grade product. DO NOT INGEST or INHALE. Keep out of the reach of children 9 & under. Non-toxic material. If powder gets in eyes, flush with water.

> Wine Dye is NOT a food grade product. DO Not INGEST or INHALE. Keep out of the reach of children 9 & under. If powder gets in eyes, flush with water. Update: Use cherry favored (sugar free) *Kool-Aid* in place of the Wine Dye.

> Gallium Spoon. Even though the gaffed spoon is made of a non-toxic alloy, once the effect is performed, the tainted coffee should be disposed of discreetly or its contents covertly recovered.

See Appendix: A, for various methods to recover the liquid metal.

Chapter 11

The Penetrating Spoon

* **Effect:** After dinner, the conversation turns to a popular discussion of life on other planets, space travel, black holes and other related topics of astrophysics.

The performer demonstrates the concept of traveling through a wormhole by using a coffee stain as a symbolic portal.

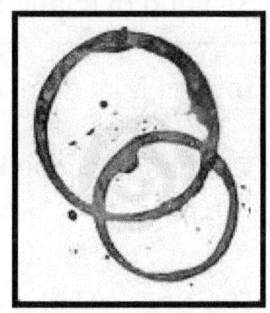

A cup of coffee is placed over the stain and a spoon is inserted into the coffee. Immediately, the spoon travels through the table-top as the performer removes it from under the table.

* **Materials Needed**: 1 gaffed spoon, 1 duplicate spoon, an available cup of hot or warm coffee and a Blu-Tack Spoon Pad.

Gallium Gaffs for Magicians

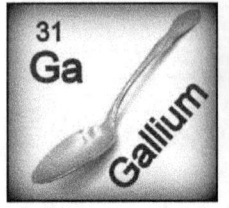   

> Option 1, gaff: Molded spoon made of Field's metal.

> Option 2, gaff: Molded spoon made of pure gallium.

> The Blu-tak Spoon Pad allows the performer to retrieve a duplicate spoon from under a table-top while in a standing position.

See end of chapter for more info on *Blu-Tack Spoon Pad*

* **Set-up:** Minimum, on-the-spot arrangement.

> The gaffed spoon is planted in place or switched out for a regular spoon.

> The Spoon Pad is covertly set-up in two movements. First, the blu-tack pad is firmly struck into place under the table-top and then after a time-delay, the duplicate spoon is securely pressed into the "putty pad". The spoon is held in place until it is needed for retrieval.

For more info on *Covert Set-up*, see Appendix: A

* **Melting-Point Method**: Immersion technique [Chapter 5].

* Time &Temperature Considerations:

Metal /Alloy	Melting Point	Coffee Serving Temp	Coffee Holding Time
Field's Metal	144°F (62°C)	200°-180°F	11 Minutes
Pure Gallium	85.57 °F (29.76°C)	200°-180°F	30 Minutes

* Presentation & Performance:

In this presentation, after pattering about worm-holes, the performer proceeds by making a coffee stain near the center section of the table.

Given that a fresh coffee strain cannot be made instantly, the stain is made symbolically by dipping the fore-finger into the coffee and making a circular pattern to represent a wormhole.

On top of the coffee stain, the performer builds a time-machine by using the coffee cup as the space capsule and the coffee as liquid fuel to propel the spacecraft through the wormhole. If available, salt and pepper shakers are placed to each side of the cup to represent rocket boosters.

While building the make-believe spacecraft, the performer engages the dinner guests to make suggestions on what other creative objects can be used to construct the vehicle.

Once done, the performer states that he will demonstrate how an object can travel through one end of a wormhole and out the other end. While in a standing position, this is done by taking a spoon and slowly pushing it into the coffee. As one hand is dipping the spoon into the cup, the other hand goes under the table-top.

When the gaff spoon melts into the coffee, the performer retrieves the duplicated spoon (held in place by the blu-tak) from under the table.

* Layers of Deception:

> As the performer talks about using an object to travel through the wormhole, he initially takes a salt shaker and as an after-thought about creating a mess, he uses a spoon – a logical object to submerge in coffee.

> Passing the spoon through the table-top while in a standing position helps eliminates the suspicion that a duplicate spoon was in the performer's lap at the time of the penetration.

> As the duplicate spoon comes into view, the performer shakes the spoon to dry it. Of course, the spoon isn't wet but the pantomime is done as a convincer.

When the spoon is moved to the hand that was just used to dip the gaffed spoon into the coffee, moisture can be transferred to give it a wet look and feel.

* **Caution:** Even though the gaffed spoon is made of a non-toxic alloy, once the effect is performed, the tainted coffee should be disposed of discreetly or its contents covertly recovered.

* **Recovery of Liquid Metal:** See Appendix: A, for various methods.

Chapter Notes

> Before the presentation, be sure to drink enough of the coffee to empty it to a half- filled level in the cup.

When the insides of the cup are viewed, the half-filled level of the coffee will dramatically confirm that the "traveling" spoon isn't hidden in the cup. It's there in another form but the eyes can't see it and the mind can't comprehend it.

> If desired, the performer can provide the duplicate spoon to match the gaffed spoon. Or, if using a generic, gaffed spoon, once the blu-tack pad is in place under the table, the performer "steals" a spoon from the available dinnerware and secures it to the pad.

In this case, when the gaffed spoon melts-away and the "stolen duplicate" is retrieved, there is no opportunity for a comparison of the two spoons utilized in the effect. The only comparison that can be made is to the actual dinnerware on the table and that would be a perfect match.

How to Make the Spoon Pad

Blu-Tack

Blu-tack (made by Bostik) is a pressure-sensitive, putty-like adhesive used to attach lightweight objects to walls or other dry surfaces. It is available in a variety of colors.

Blu-Tak Spoon Pad

This unit is constructed by embedding a duplicate spoon into a pad of blu-tack. The pad is covertly stuck in place under a table-top and then the back-side of the spoon's bowl is pressed upward into the putty. This positioning allows the handle to angle downward for an easy-to-find grip when removing the spoon.

The LOCTITE FUN-TAK version of the putty comes in five, 4.5 inch strips of putty.

> Take three of the strips, place them side by side and knead the middle seams, sides, and ends together to make one unit.

> Then take this unit and evenly fold it half. Press the halves together. This will give the pad a good thickness (0.25 inches). The overall pad is

approximately 1.75 by 2 inches – enough to securely cover the bowl surface of the spoon.

Tip: Save the wax paper that comes with the putty because you can use it to maintain your pad when storing or transporting it.

Chapter 12

The Penetrating & Restored Spoon

* **Effect:** In this version of the *Penetrating Spoon,* the spoon partially travels through the table-top in the first attempt, where the performer ends up with the bottom section of the spoon (bowl) and the spectator is left holding just the upper part (handle). Finally, as the spectator re-inserts his portion of the spoon into the coffee, the performer reaches under the table with his piece and then removes the whole spoon "restored".

* **Materials Needed:** One gaffed spoon, an available cup of hot or warm coffee and a Spoon Caddy [containing a partial spoon & a whole duplicate spoon]

  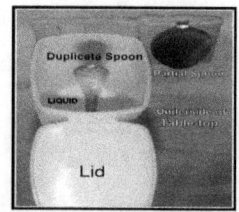

> Option 1, gaff: Molded spoon made of Field's metal.

> Option 2, gaff: Molded spoon made of pure gallium.

> The Spoon Caddy will enable the performer to produce a partial spoon and then a whole spoon dripping wet.

For info on building the *Spoon Caddy*, see end of chapter

* **Setup:** Maximum -- Prior arrangement needed to install the Spoon Caddy under the table-top.

* **Melting-Point Method**: Immersion technique [Chapter 5].

* **Time &Temperature Considerations:**

Metal /Alloy	Melting Point	Coffee Serving Temp	Coffee Holding Time
Field's Metal	144°F (62°C)	200°-180°F	11 Minutes
Pure Gallium	85.57 °F (29.76°C)	200°-180°F	30 Minutes

* **Presentation & Performance:**

This effect is presented in the same manner as the *Penetrating Spoon*, where the coffee stain (wormhole) is drawn upon the table and an imaginary spacecraft (coffee cup) is built with the available dinnerware on the table.

In this version of the trick, the spectator is allowed to submerge the gaffed spoon into the coffee.

The spectator is instructed to dip the spoon half-way into the coffee and after stirring; he is instructed to remove it. Upon removing the spoon, the spectator will only have the handle of the spoon because the bowl of the spoon will have melted away.

At this point, the performer reaches under the table and produces the bottom piece of the spoon. The performer's piece of spoon is now displayed alongside the spectator's portion.

Next, the spectator is instructed to drop his portion of the spoon into the coffee and then to use another utensil to stir through the liquid to verify that there is nothing at the bottom. At this moment, the performer reaches under the table with his half of the spoon and dramatically removes it fully restored.

During this procedure under the table, the performer places the half-spoon back onto the side-pad and then carefully removes the regular spoon from its container with the bowl of the spoon upright and full of liquid. As the spoon is brought to the table top, the performer shakes the wet spoon dry.

* **Caution:** Even though the gaffed spoon is made of a non-toxic alloy, once the effect is performed, the tainted coffee should be disposed of discreetly or its contents covertly recovered.

* **Recovery of Liquid Metal:** See Appendix: A

Chapter Notes

> When the duplicate spoon is removed, it is held horizontally

under the table-top to ensure that the bowl's liquid will not spill. As the spoon clears the edge of the table, the wrist turns it upright causing the bowl's contents to spill over the hand --- a nice convincer that the wet spoon just traveled through the coffee cup and table-top.

How to Build the Spoon Caddy

The caddy consists of two sections:

1. The spoon- pad. This is the holding area for the partial spoon. The end of chapter 11 contains details on how to make the spoon-pad with blu-tack putty.

2. The wet-container. This is the section that houses the duplicate spoon. Once the container is leveled with water or coffee, the duplicate spoon can be retrieved wet.

The actual container in the photo is a plastic candy box with a lid. Any container can be utilized. It must have the proper width to clear the bowl, good depth to hold an ounce of water (or coffee) and the right length to hold the spoon in place.

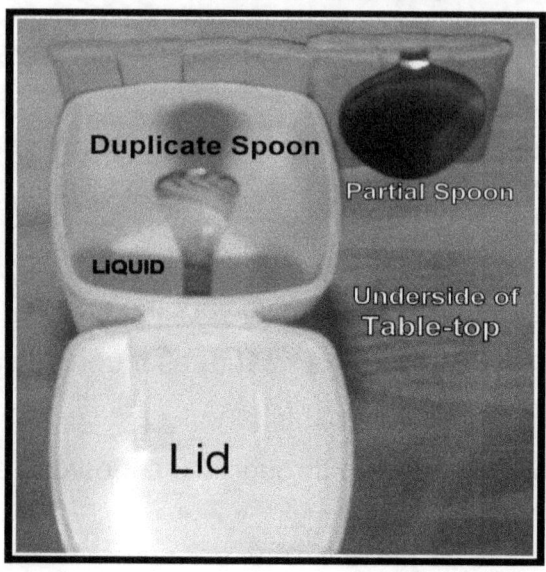

Once an ideal container is selected, it is placed on top of a pad of putty. Blu-tack and other versions of the putty come in 5 strips (4.5" long) that can be cut and shaped into a mat to hold the container.

The container in the photo required 4 strips. The strips were laid side-by-side and then the middle seams, sides, and ends were kneaded together to make the unit solid.

Depending on the shape of your container, it may take a small strip of putty to build a slight incline to tilt the back-end downward. This is done to ensure that the added water will cover the bowl of the spoon and level without spilling forth.

Once the wet-container is made, the spoon-pad is placed to the right or left side of it. Both pads are kneaded together to form one unit.

Making the Partial Spoon

The partial spoon is made by taking a regular spoon and

cutting off the bowl / neck area. This is easily done by sawing or filing. Once a cut is notched, the spoon can be bent back and forth to make the break.

The photo shows a partial spoon made with soldering wire to give it a freshly melted look.

Chapter 13

Tuning into a Shrinking Spoon

* **Effect:** While preforming at the local coffee shop, the performer presents a magical tuning fork, claiming that it has some special functions beyond tuning musical instruments.

In a demonstration, the tuning fork is set in motion and tapped against a borrowed spoon. Suddenly, the spoon shrinks to half its size.

 * **Materials Needed:** An available cup of hot or warm coffee, one gaffed spoon, a matching short spoon and a tuning fork.

> Option 1, gaff: Molded spoon made of Field's metal.

> Option 2, gaff: Molded spoon made of pure gallium.

  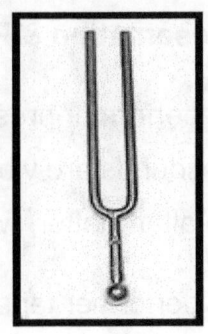

* **Setup:** Minimum -- On-the-spot arrangement.

Prior to the performance, the short spoon is covertly placed in a cup of coffee, submerged and hidden under the coffee's surface.

During the demonstration, the gaffed spoon is shown and held with the handle protruding out of the coffee while the spectator sets the tuning fork in motion. As the spoon is touched by the tuning fork, it is released to completely melt-down and the short spoon is removed from the cup, implying that it has shrunk in size.

* **Melting-Point Method**: Immersion technique [Chapter 5].

* **Time &Temperature Considerations:**

Metal /Alloy	Melting Point	Coffee Serving Temp	Coffee Holding Time
Field's Metal	144°F (62°C)	200°-180°F	11 Minutes
Pure Gallium	85.57 °F (29.76°C)	200°-180°F	30+ Minutes

* Presentation & Performance:

The performer presents a magical tuning fork and explains that sounds are vibrations that move through matter but can also alter matter, which will be demonstrated shortly.

The performer taps the tuning fork, holds it near his ear and patters:

"When a regular tuning fork is struck, you cannot see the sound waves travel out from the fork. Yet, you can hear them because the air molecules bounce off the fork, moving through the air until they reach your ear-drum. The movements of air molecules are only visible if the vibration occurs in a denser medium such as water or a liquid like coffee".

The performer strikes the tuning fork again and holds it near the surface of the coffee or slightly dips it into the liquid.

"When a common tuning fork is set in motion, the patterns of the vibrations can be seen in the disturbance of the water. What's really interesting is that when a magical tuning fork is tapped against something metallic and a little heat is applied, the molecular properties of the material can be altered".

The performer proceeds by borrowing a spoon [actually the gaff spoon, planted in advance] and placing it half submerged in the warmth of the coffee while instructing the spectator to set the tuning fork in motion. The tuning fork is touched against the spoon and as the "vibrating spoon" is removed, it has shrunk to half its size.

Layers of Deception

> The entire presentation is about the tuning fork.

> The gaffed spoon is planted in advance. There's nothing to suspect because the spoon is "borrowed" from its natural habitat (dining table).

> When the performer submerges the spoon, he keeps the top half of it sticking out of the coffee while the spectator is instructed to vibrate the tuning fork and then to dip it into the liquid. As the coffee is displaced and splattered (due to the vibrations) the performer lets the protruding part of the spoon sink into the cup.

This procedure is done to visually emphasis that the tuning fork is energizing the water and thus the spoon is being transformed in the process.

* **Caution:** Even though the gaffed spoon is made of a non-toxic alloy, once the effect is performed, the tainted coffee should be disposed of discreetly or its contents covertly recovered.

* **Recovery of Liquid Metal:** See Appendix: A, for various methods.

Chapter Notes

The requirements of the short-spoon are:

> It should be far smaller in length and shape than the gaffed spoon. Also, it must match in style. The size difference of the gaffed spoon going into the coffee and the short spoon coming out of the cup should be easily noticeable. Basically, if your gaffed spoon is 6 inches in length, then your small spoon would be 3 inches or less.

> It must also be short enough to hide completely in a cup or mug containing coffee. Most coffee mugs or foam cups are suitable for hiding a smaller spoon.

Should the mug barely hide the spoon, you can add more coffee or bend the spoon's bowl and handle into an "L" shape to get the proper coverage. During the performance, when the short spoon is being retrieved, unbend it as you remove it from the mug. Act as if you're squeezing it to shrink the size.

Options for making a short spoon:

1. The naturally small size of a demitasse spoon is ideal for minimal modification. For even smaller spoons that need no modification, use miniature cocktail spoons.

The demitasse / mocha spoon is approximately 3¾ to 4½ inches long. It is used in formal dining where coffee is served in a demitasse cup and a demitasse spoon is used if sugar is added.

2. Purchase a plain- style baby spoon that approximates the style of the gallium spoon. You can find a baby spoon that is a near match or one that can be easily modified to make it so with a little grinding and sanding applied to any embellishments.

Chapter 14

When the sun dies and the stars fade, a voice will be heard to say: "What… another card trick"? --- Ruminations of Riku, Sora & Hero

Spoon Boy Prediction

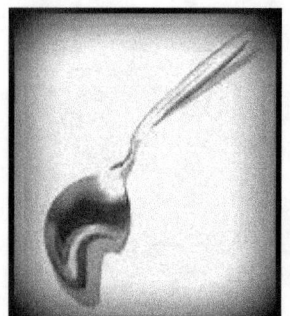

* **Effect:** The performer presents a prediction made by the "spoon boy" of the *MATRIX* movie. A spectator is requested to privately select a card and concentrate on it. When the prediction is opened, it reads: *Concentrate on the Spoon*.

Sensing that the message means something, the spectator is asked to concentrate on the words. Mindfully, the writing visually turns into symbols that represent the correct card. Not to be outdone, the performer demonstrates his power of concentration by bending and melting a spoon in honor of the spoon boy.

* **Materials Needed:**

> Alloy: A molded spoon made of gallium and indium [#3].

> A FRIXION pen and gaffed prediction note (Appendix: A). A *Blaze Craze Tuck Case* (chapter 23).

> A regular deck of cards, a few paper clips and a plastic bag.

* **Setup:** Maximum setup. In this effect, the props are best managed from a prop-box or personal suitcase. The gaffed spoon must be brought into play from a *spoon ice-bag* (chapter 23).

* **Melting-Point Method**: Exposure Technique [Chapter 5].

* **Time &Temperature Considerations:**

Metal /Alloy	Melting Point	Exposure Technique	Melt-Rate
#3 Ga+In	60°F (15.7°C)	X	* Rapid*

* **Presentation & Performance:**

Briefcase Setup: The case contains the gaffed spoon (housed in the ice bag), a loosely banded deck of cards and a plastic bag containing the *Blaze Craze Tuck Case,* a few paper clips, and the gaffed prediction note (written on a folded index card).

> During the dinner-time conversation, the gaffed spoon is causally bought into play by placing it into a glass of ice water or tea. This setup allows the spoon to be in view during the entire performance and it's protected from premature melting.

Phase 1: Select a Card

The performer starts by announcing that he has a special prediction made by the spoon boy of the Matrix movie. The plastic bag is brought forth and the contents are removed. The gaffed tuck case (*Blaze Craze Deck*) is momentarily placed in the briefcase to free the hands, while the prediction note is sealed with a paperclip. The plastic bag is tabled for later use.

> As the spectator is handed the prediction for safe-keeping, the performer gets rid of the extra paperclips by placing them into the briefcase and removing the uncased playing cards with along with gaffed tuck case. From the spectator's point of view, it's assumed that the cards were uncased in the briefcase while organizing the paperclips.

> *Note*: The BCD is turned on when exiting the briefcase.

The performer tables the tuck case and holds onto the cards asking a spectator to make a selection.

> The 8-of-spades is forced on the spectator by any method preferred. The forced card is retuned and the deck is tabled.

Finally, the prediction note is opened to reveal the words, CONCENTRATE on the SPOON. Sensing that the message means something, the performer looks towards the glass of ice water and removes the gaffed spoon, checking to see whether it has a message engraved on it or maybe he can use it as a divining rod to find the selected card.

> *Note:* During all times, the spoon should be held by gripping the edges of the handle to prevent any possible stains due to transfer of body heat.

> Before the spoon is checked for clues, the performer places the note on top of the gaffed tuck-case to free the hands. Care is taken to ensure that the message part of the note isn't placed directly over the "hot spot" on the tuck case.

After more contemplation regarding the spoon, the performer tables it (allowing it to continue cooling down for phase-2) and in a moment of inspiration the performer patters:

"It's not the actual spoon that should be concentrated upon, it's the <u>underlined</u> spoon. It's the spoon written on the prediction".

The performer picks-up the note and tuck case, requesting that the spectator concentrates on the note.

> The rationale for picking up the tuck case is that it's used as a mini clipboard. Also, the performer gestures between the prediction note and tuck case when telling the spectator to concentrate on the note and not the cards that came out of the deck (tuck case).

> When the note's message is centered over the hot-spot on the tuck case, all the letters and underlining will vanish except the "S" and

105

to two "OO"s. Here, the performer smiles and states that the double "OO"s, side-by-side represent an "8" and the "S", means a spade.

Phase 2: Spoon Bend and Melt

When the prediction is verified as correct, the performer states that he is going to proudly frame the note and spoon as souvenirs.

> The spoon is placed in the plastic bag as if it were a trophy for safe-keeping. Actually, it's placed in the bag to collect any degree of the melting when it's placed on the tuck case.

> *Note:* The bowl and neck section of the spoon is placed directly on top of the BCD's hot spot.

As the bagged spoon is placed on the tuck case (to be thawed), the performer creates a further time delay by picking up the playing cards to find the 8-of-spades as part of the souvenir collection.

During this time delay, the performer discretely monitors the condition of the spoon. It will take 3-5 minutes for the spoon to thaw enough to have a visual display.

The following options are available:

> If the spoon is in a good state of melt-down, then direct attention to the display. To speed things up, the performer can

place his finger upon the spoon's neck to slowly bend it downward while it remains on the tuck case.

> If the spoon is in a slow state of melt-down, then pick it up by pinching the neck between the forefinger and thumb. With a little pressure, the spoon can be made to bend and then continue melting.

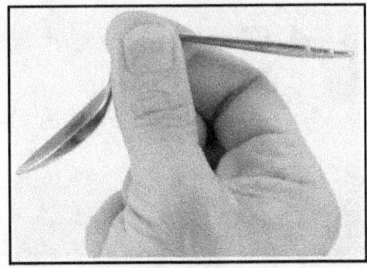

You just out performed the spoon boy --- you mentally bent and melted a spoon.

* **Caution:** No caution for the melting liquid because it's conveniently collected in the plastic bag. Caution should be taken to ensure that the *Blaze Craze deck* is turned off after it is used.

* **Recovery of Liquid Metal:** Self-contained.

Chapter 15

Effect: Mental Melt Medallion

* **Effect:** The performer presents a medallion and tells an interesting story about it. In a demonstration of mind over matter, the performer causes it to visually melt in his hands.

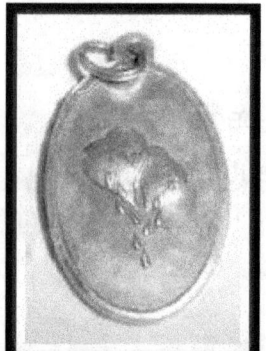

* **Materials Needed:**

> A molded, copied medallion made of gallium-indium [#3] – casted from a regular medallion (original).

> A plastic *Zip-Lock* storage bag containing the regular medallion, a marker and some miscellaneous items (coins, paper clips).

* **Setup:** Maximum. The gaffed medallion must be kept chilled (chapter 23) until bringing it into play.

* **Melting-Point Method**: Exposure Technique [Chapter 5].

* **Time &Temperature Considerations:**

Metal /Alloy	Melting Point	Exposure Technique	Melt-Rate
#3 Ga+In	60°F (15.7°C)	X	* Rapid*

* **Presentation & Performance:**

The story-line that you tell about the medallion will be based on the actual medallion that you use to mold and cast the gaff. Based on the theme of your medallion, the tale that you tell should end with the justification to magically melt the medallion.

With a bleeding-heart medallion, the patter is based on the theme that hearts engraved in silver will shed silver blood as represented by the melt-down of the medallion.

From the start, the regular medallion is removed from the plastic storage bag and presented to the spectators. They are allowed to handle it while you are telling your story. Here, there is nothing to prove, you just have a showpiece.

Near the climax of your story, you empty the remaining contents of the bag into your hands. The spectator is handed the marker along with the empty bag.

At this point the suggested patter is:

"I want you to do me a favor and sign this plastic bag across the top section with your full name. I know that this is a little strange but you'll see why in a moment."

As you give the bag and marker to the spectator, you retrieve the regular medallion --- you are taking it to free up their hands.

As they are marking the bag, you simply put the medallion in your pocket and switch it for the gaffed medallion.

> When working close-up, the gaff is retrieved from your pants or jacket pocket. Prior to preforming, it is placed in a mini-ice bag to shield it against prematurely melting. See chapter 23 for details.

This procedure is not a move; it's just you placing the miscellaneous items from your hands into your pockets. If they see you do this, it's ok because you are organizing things.

When the signing is done, you take the bag and immediately place the gaffed medallion into it.

> * Once the gaffed medallion is brought forth from its ice bag, you have about 30 seconds to place it in the plastic bag before any noticeable staining occurs. To minimize any heat-transference, the coin is held at the edges before placing it into the plastic bag.

The medallion is placed into the center of the bag and pinched in place between the thumbs and forefingers on both sides. The fingers should cover as much of the medallion as possible without blocking the view of its center section.

* This is the time that you tightly massage the medallion with your fingertips to transfer as much body heat as possible. The plastic bag should be slightly squeezed inwards to create a hollow that will allow the melting drops to fall free. If the plastic bag is flattened against the sides of the medallion, the visual flow of the melting process will be restricted.

As a minute approaches, the medallion will start to visually melt. At first, a liquid droplet will form and then break free as the metal progressively melts.

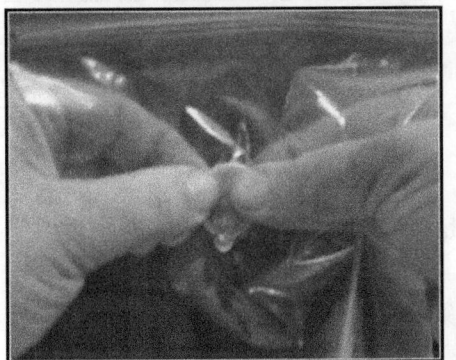
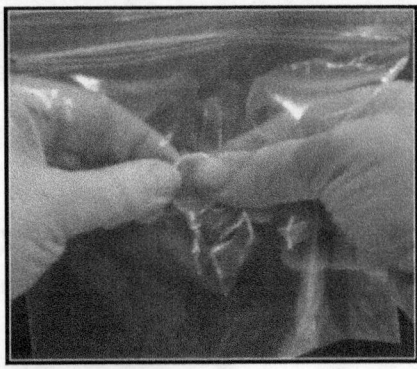
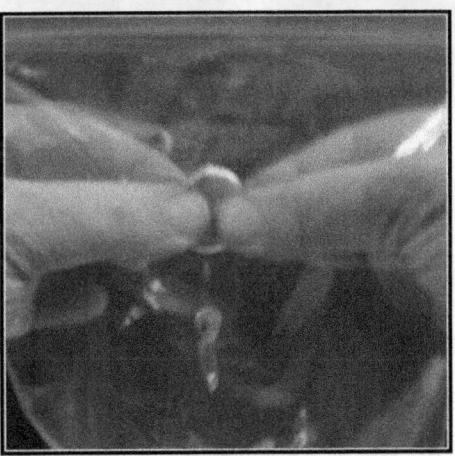

At this moment, it's all showmanship to dramatize the effect. If preferred, you can make patter about how the thoughts you are directing towards the medallion are *"cool, calm and collected"* and then hand over the melting medallion (still inside the bag) to the spectator to hold in the same manner that you held it for continued melting.

Here, the spectator will experience a contradictory feeling – the melting medallion is not hot but cool to the touch --- "cold blooded tears of silver".

* When a gallium-indium alloy melts, it's still cold to the touch during the process, a rather "cool" effect to capitalize upon.

* Layers of Deceptions

> From the start of the presentation, the storage bag and its contents are brought into play. The plastic bag is justified because it safely holds the unfastened medallion which is missing its chain.

> As you tell the story, the spectators are allowed to handle the normal medallion to illustrate what you are talking about. In this situation you are not "over proving" but you are indirectly suggesting that all is real, which it is.

> The signing of the bag is two-fold. It allows the misdirection for switching the medallion and it implies that no switch can be made. That's what we magicians do; we dare to lie in words and actions to make delightful deceptions.

* **Caution:** None. The melting liquid is conveniently collected in the plastic bag.

* **Recovery of Liquid Metal:** Self-contained.

Molding Options

You can purchase the desired medallion (original) and make a mold of it to cast / copy your gallium-indium gaff.

The metallic color of the master medallion should approximately match the chrome color of gallium-indium. It doesn't have to be a perfect match given that there is never a side-by-side comparison of the two medallions.

The back-sides of medallions are flat, which is easy to copy with an open-top type of silicone mold.

> See chapter 22 for DIY Molds

Super Easy Mold Based on a Blank Representation

If you want to test this effect beforehand, you can use a compound such as clay, *Play-Doh* or even *Silly-Putty* for making a quick and easy mold.

These types of compounds don't create good details but are good for experimentation purposes without going through the total expense of making a mold.

> The compound is shaped into a small mound and then impressed with a quarter-size coin, disc, or even the back-side of your original medallion.

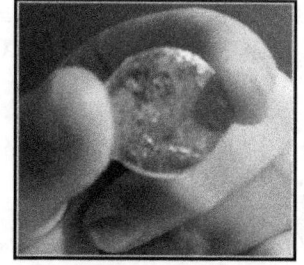

> Once the indentation is made, you can pour liquid gallium-indium into the mold. When the metal solidifies, you'll have a double-sided blank representation of a medallion to test.

Coin Mold

You can do this effect with a silver-chrome type of coin but it will have to be one that you present because there will be very little opportunity to borrow a matching or near-matching coin in terms of metallic color. Gallium-indium has a silvery shine in contrast to the clad [copper-nickel mixture] color of U.S. coins.

> Even though the regular coin and the gaff are never presented in the same space and time for a side-by-side comparison to be made, the shade-matching precaution is for that one eagle-eyed or expert spectator who may notice a difference.

An India 50 Paise coin is a closer shade to the color of gallium-indium than is a U.S. quarter. Also, the unfamiliarity of foreign coins or tokens allows for any slight differences in color to go unnoticed.

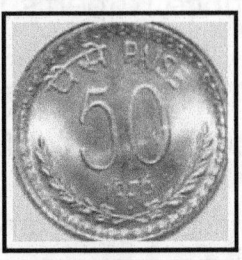

When making a coin mold, the only practical mold that can be achieved is an open-top mold. This type of mold will produce a coin showing only one side (heads or tails) of the coin's profile and the opposite side will be blank. This limitation is adjusted for in the routine by showing only the detailed side of the gaffed coin while holding it within the plastic bag from the spectator's point of view.

A *silicone putty* mold is ideal for making molds of the older coins that have a raised, high-definition profile. In order to capture the weak details of the newer, lightly stamped coins, you'll need a low viscosity, *liquid silicone rubber* that will reproduce the finest details. This will also have to be an open-top mold for the average hobbyist.

Frozen Plus Melted

Anyone acquainted with the trick FROZEN (by Adam Grace) may realize that the "napkin method" is ideally suited for temperature protecting gallium-indium gaffs. The effect is extended by taking the frozen coin and then melting it.

I'm not promoting any products here, just offering options for the thoughtful magician to explore.

Prince Rupert's Drops

What does a "glass-tadpole" have to do with a chapter on gallium-indium gaffs? Read on.

Prince Rupert's drops (aka: Dutch tears) are glass droplets 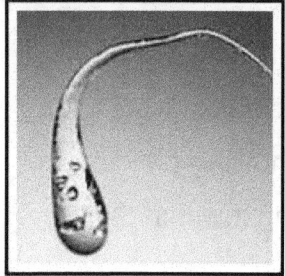 formed by dripping molten glass into a bucket of water. When a drop of melted glass comes in contact with cold water it instantly cools into a tadpole-shaped droplet that has a long, thin tail.

Interestingly, the head section of the body can be hit by a hammer and the droplet will not break but when the tip of the tail is snapped, the whole structure explodes, shattering into gain-size pieces of glass.

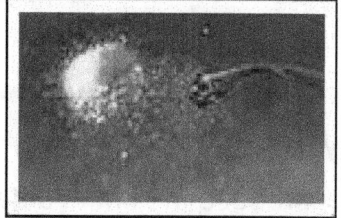

This glass oddity has an application as a mind-over-matter type trick. In performance, the Rupert's drop is presented with the theme of a "tear-drop frozen in time" or as a "tears hardened with grief". It is demonstrated as indestructible by hitting the head section with a hammer.

For eye protection, the drop is placed in a plastic bag. To cause the "mental explosion", the performer simply pinches the tail. The technique is to hold the tail between the thumb and index-finger. In this position, the middle finger is used to snap the tail. This is easily done without any perceptible movement of the fingers.

> For more information on *Rupert's drop*, see Appendix: A

Two bags of Weirdness

The *Mental Melt Medallion* can be combined into an extended routine by using the Rupert's drop effect. The themes of bleeding heart (blood drops) and glass tears can be woven into an integrated story-line.

In preparation for the performance, three or four Rupert's drops are placed in a small plastic bag. This bag is then placed in the larger bag that contains the medallion and miscellaneous items.

During the original routine, when the "medallion bag" is being emptied, the "Rupert bag" is handed to a spectator for safe keeping. Then the routine proceeds as outlined.

> For consistency of actions, at the time the medallion bag is signed, the bag with the Rupert's drops is also signed.

When the medallion-melting is completed, the Rupert's drop phase of the routine starts. In the end, the spectators are left with two bags of weirdness to ponder about.

Chapter 16

Hitched Ring and String

* Effect: An interesting story is told while a string is knotted on a ring by an audience member. Within seconds, the magician has freed the ring while the spectator holds the ends of the string.

* Materials Needed:

> LMP Alloy: A molded, gaffed ring made of Field's metal. The inside of the ring is engraved with: Fat Free Coffee Mate.

> A long string or shoe lace and a regular, matching ring with the same engraving.

> An available cup of hot coffee.

* Setup: On the spot. Prior to the performance, the matching, non-gaffed ring is covertly planted in the cup of hot coffee.

* **Melting-Point Method**: Immersion technique [Chapter 5] or the flame from a match for the bar bet version of the trick.

* **Time &Temperature Considerations:**

Metal /Alloy	Melting Point	Coffee Serving Temp	Coffee Holding Time
Field's Metal	144°F (62°C)	200°-180°F	11 Minutes

* **Presentation & Performance:**

This trick can be presented as a magic effect or as a bar bet. See end of chapter for details.

The basic routine is to present the spectator with a ring and string, while giving him instructions on how to tie a simple Lanyard hitch to secure the ring.

Performance Patter:

"The Lanyard knot is the best hitch knot used when attaching a rope to an object. The tie has been known since the first century and was described by the Greek physician Heraklas as a useful surgical knot.

Throughout the ages the knot has been know under a variety of names: Cow hitch, Lark's foot, Ring hitch, Deadeye hitch and Coffee's Mate.

During the 15th century, a French businessman and sailor by the name of Coffee perfected variations on the hitch knot while sailing the seas trafficking in vanilla beans.

Beyond being a successful businessman of his day, Coffee was famously known as a master of escape on land and at sea. He could slip any knots tied to his wrist and body. This was an amazing feat because Coffee was a rather large man in girth. He tipped the ship's scales at 400 pounds. You can say, he was the "big" predecessor to Houdini".

At this point, as an inspired pun, the performer uses a cup of coffee in the spirit of Coffee the man and states that he will free the ring without touching it.

As the tied ring is about to be lowered into the coffee, the performer patters:

"Did you notice the engraving on the inside of the ring? No? That's ok; we'll get to it later".

The tied ring is submerged into the hot coffee as the spectator securely holds the ends of the string.

The performer taps the sides of the cup (ship's bell) three times with a spoon to summons the essence of the long dead businessman, escapist, and shipmate.

When the performer is sure that the gaffed ring has melted to the bottom of the cup, he instructs the spectator to slowly lift the bare string revealing no knot and no ring.

The performer removes the planted ring as if it were the freed ring and then requests the spectator to read the engraving on the inner-side of the ring. It reads: Fat Free Coffee Mate.

Layers of Deceptions

> By any definition, the gaffed ring is the real thing. It's a ring that has been casted with a molten metal, which happens to be Field's metal. No amount of inspection will reveal a gimmick.

> The story-line is told to establish a verbal and mental connection between the gaffed ring and the duplicate being perceived as the same ring without the possibility of a switch given the engraving.

> The rationale for the use of the coffee is justified by the story-line, the puns, and the ring's comical engraving.

* **Caution:** Even though the gaffed ring is made of a non-toxic alloy, once the effect is performed, the tainted coffee should be disposed of discreetly or its contents covertly recovered.

* **Recovery of Liquid Metal:** For various recovery methods, see Appendix: A.

Options for Engraving the Rings

Option #1: Do-It-Yourself Ring-Mold.

> A regular, existing ring is engraved on the inside and used as the master ring to make the ring-mold. This ensures that the gaffed ring made of Field's metal will automatically cast with the same inscription.

Note: The master ring that you start with must match or near-match the metallic color of Field's metal.

Option #2: Purchasing a Ready-Made Ring Mold.

If you don't want to make your own mold, you can purchase an off-the-self ring mold. In this situation, you have a mold but no rings. You'll need an original ring and a gaffed ring.

> The first step is to cast the original ring – this is the one that will be hidden in the cup of coffee (until the gaffed ring is used). You'll have to make the cast with any LMP alloy that has a melting point above 212°F and under 395°F (the tolerance level for most silicone molds).

Note: A mixture of 58% bismuth and 42% tin has a melting point of 281°F, which is a non-toxic, easy to cast alloy.

Since coffee is served between 200-180°F, the original ring will be able to withstand the heat when hidden within the cup.

Once both rings are casted, they are engraved. The composition of Field's metal is strong enough to be engraved upon.

See chapter 21 for Ready-Made Ring Molds & chapter 22 for DIY Molds

Open- Top Mold for Washer Substitute

This effect can be done with bolt-washers in place of the rings. Washers come in all sizes and metallic shades that will match or near match Field's metal.

Since washers are two-sided flat objects, an open-top type mold made of *silicone-putty* is a quick and easy option for casting the props.

See chapter 22 for more Information on DIY Open-Top Molds

Bar Bet Blow-out

This effect can be done as a bar bet where a wager is made that you can free the tied ring with only a match.

The spectator will immediately assume that you will burn the string to untie the ring. At this moment, you light the match and blow it out while stating that you can do it without touching, burning or destroying the <u>string</u>.

Once the rules are established, you have the spectator hold the ends of the string and you re-light the burnt match, holding it against the dangling ring. The heat from the flame will melt an opening in the ring, providing the means to free it.

* The match that was lit and blown-out was switched for a gaffed match. The gaffed match was made by taking a regular match and blacking the head to give it the appearance of being burnt.

If you want to make a truly realistic looking burnt match, see *Re-Light-able Matches*, Appendix: A.

Chapter 17

Coke Can Crush & Crumble

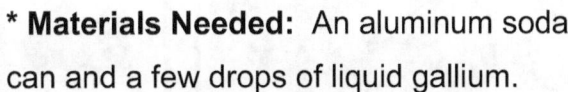

* **Effect:** A soda-can is not merely crushed with bare hands but is made to crumble to bits and pieces as if it were made of paper mâché.

* **Materials Needed:** An aluminum soda can and a few drops of liquid gallium.

* **Setup:** Minimum. The prepared soda canister is planted in a trash-can or placed with other public litter.

* **Method**: A small dab of liquid gallium placed on any aluminum object will induce some degree of structure failure through a phenomenon known as *liquid metal embrittlement*.

Gallium easily diffuses with many metals but when it reacts with aluminum, it produces an alloy which is very brittle.

* **Presentation & Performance:** This effect can be presented as a demonstration of mind-over-matter or as a feat of super-human strength.

* **Caution:** Once the canister is crushed, it should be properly disposed of for the reason that the gallium may still be reacting to the aluminum. Do not attempt the preparation with a sealed soda-can. The moment that the gallium effectively weakens the canister, the pressurized contents will erupt.

Chapter Notes

To get the full crumbling effect:

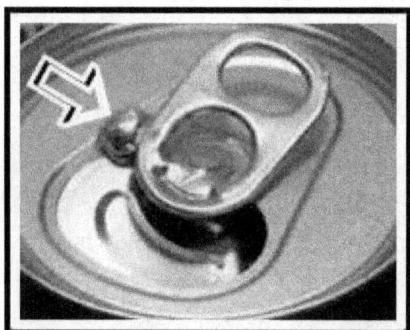
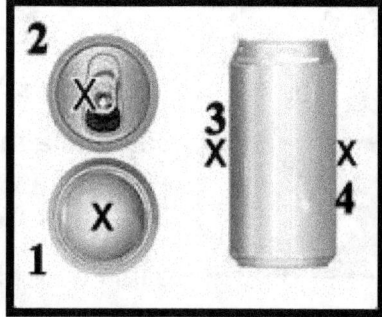

Step-1: Start by turning the canister upside-down and place a small drop of liquid gallium onto the center of the can's bottom.

This is best done with a syringe. Once the drop is settled, take an awl or needle and poke through the drop while scratching the surface beneath it. The scratching removes any layer of oxidation, allowing the gallium to fuse with the aluminum quicker. Within 30 minutes the gallium will compromise the aluminum.

Step-2: You can now upright the canister to place a drop of gallium onto the top surface. Again, scratch the surface and wait 30 minutes for the gallium to fuse and spread.

Step-3: Turn the canister on its side and make a small dimple with your thumb. Situate a drop of gallium into the depression and do the scratching procedure as before. Wait 30 minutes for the gallium to fuse and spread.

> *Note:* The slight indentation should be made as a "smooth curvature". If you press to hard, you'll create a sharp-angled dent and as the fusing spreads, the canister will crack with a popping sound at that point. If this happens, re-apply a new gallium droplet.

Step-4: Do the same as in step-3 but on the other side of the canister.

Each drop of gallium will leave a corrosive spot on the canister. This situation is still presentable because a soda-can retrieved from the trash would have a certain "beat-up" look.

> *Note:* During this process, the gallium droplets will become contaminated with aluminum; therefore do not consolidate the mixture with a pure gallium source.

Chapter 18

Utility: Gambler's Gallium

Whenever gallium is exposed to air, a thin layer of gallium oxide forms on its surface, causing it to "wet" almost any material. This natural process makes it ideal for making brilliant mirrors and coating objects to make them conductive.

To make a do-it-yourself mirror, take a cotton swab and dip it into a vial of gallium, twirling it around to get a good coat. Then evenly rub the coated swab on a glass surface. The liquid metal will stick to the glass making a reflective layer as seen through the glass.

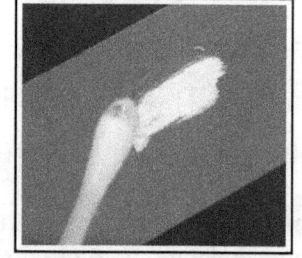

Furthermore, using this technique with a glass cutter and glass microscope slides, you can custom-make your own shiners in various sizes and shapes.

Shiners (also known as glimmers) are any reflective surface that a card cheats uses to peek at cards as they are being dealt. Magicians and mentalist use shiners to covertly peek the identity of a chosen card during a performance.

Shiner Variations

* Blackjack Prism – The prism of glass is situated in a blackjack shoe to reflect the identity of the next card to be dealt. The prism works like a miniature periscope.

* Tableware – Well polished knives, spoons and forks are reflective enough to mirror a card's index.

* Money Clip – A wad of money held with a money clip can be used to identify a card when the dealer directs the deal over it.

* Coffee – Under strong lighting, the surface of black coffee in a cup will reflect a card's identity.

* Cygnet Ring – A very common "hand-held" shiner is any ring with a polished flat surface.

* Smoking Pipe --- A small, circular mirror placed into the bowl of a smoker's pipe can be used as an effective shiner. The pipe is placed on the table and titled towards the dealer to view the cards being dealt.

* Cigarette – A polished thumb-tack or push-pin inserted into the filter end of a cigarette can be used as a shiner. The

cigarette is held between the first and second fingers of the dealing hand to view the cards as they are dealt.

* Mobile Phone – The black face (off-mode) of a mobile phone is reflective enough to shine a card's identity. Also, there are numerous smart-phone apps that will display a black screen while actively videotaping. Any cards dealt over the "fake screen" can be view in the camera roll, under the pretext of answering or making a call.

Chapter Notes

Porcelain is used to make household wares, decorative items and art objects amongst other things. Given that gallium will wet porcelain, this can be a further source for making shiners without the use of glass.

Chapter 19

Gallium Grid Marking

Gallium staining is an outstanding **daub** material, probably the best for an impossible-card-location effect. It's a secret that I've kept for years because it has enabled me to "fool the best minds in magic" as the saying goes.

Playing the Daub

Daub is paint -- any pasty colored substance (usually red or blue in color) that is utilized as a method to mark cards while in play. The substance is made out of numerous materials including blue or green eye-shadow, all types of grease, soapstone, lip stick, pencil graphite and even cigarette ash.

When secretly marking cards, the magician or cheat smears the targeted card with daub, leaving just enough of a smudge on the card to identify it when needed. The smudge is applied with the tip of the thumb or index-finger on the back-side of the card's design.

Magicians mainly use daub as a method to "tag and track" a spectator's freely selected card. As the card is being returned to the deck, the magician tags it with daub in order to identify it at a later moment.

Interestingly, when you apply gallium stains to the white body  area of the bicycle-riding angel, the smear will expose the texture of the card's white grid-type pattern. The pattern is very subtle but very obvious when you know what to look for.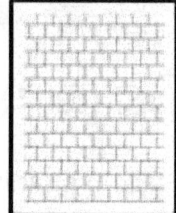

Gallium daub is excellent the following reasons:

* It's not greasy, sticky or slick when applied.

* It doesn't smear onto the other cards in the deck when applied properly.

* The "work" can be removed from the card. The stain can be wiped clean from the card with a swipe of the fingertip just as easily as it was applied.

* No detectable shine. One way to detect daub smears is to angle the card towards a light source to identify any reflective shine as evidence. Gallium stains won't create tell-tale shining.

Limitations:

* The gallium daub is best applied to blue-backed cards.

If the daub is slightly over applied, the card's blue background will help camouflage the excess amount. It can be used on red cards but the gallium stain is clearly seen as a gray smear.

* As stated earlier, this daub is perfect for an impossible-card-location trick. Yet, it won't easily work in a gambling environment because its repeated usage will invariably lead to some type of accidental smearing. Once a minuscule amount of gallium smears, it really smears.

The flesh of the skin easily cleans any smearing with just a wipe of the finger. If you try to clean the smear with a cloth or paper towel, it will act more like a brush and spread the smear rather than remove it.

How to Make & Use a Daub Coin

The best way to handle gallium as a daub is to form it into a wafer or disc shape.

This is done by taking a modeling compound such as clay, *Play-Doh* or even *Silly-Putty* and making a small supporting mound. The compound is then impressed with a penny or small bolt-washer.

Once the indentation is made, you can pour liquid gallium into the mold. When the gallium solidifies, you'll have a daub-coin which can be stored in any small container for pocket management.

Super Easy Mold

An even easier mold and container to use is a *contact lens case*. Just pour the gallium into one side of the container and let it solidify.

Don't remove the gallium -- let it stay nested in the slot and cap it. Cut the case in half to make an ideal storage container.

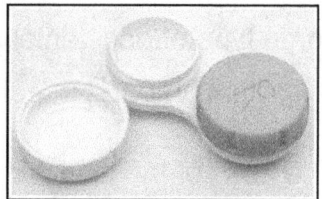

When you need the gallium, remove the lid and pinch the sides of the open container between your index-finger and thumb to receive a stain for daubing. This procedure is easily manage with one hand in your coat or pants pocket.

Chapter 20

Goofing with Gallium

This chapter contains pranks, ponderings, and other projects with gallium.

The Dissolving Spoon Prank

In this effect, two Alka-Seltzer tablets are placed in a warm glass of water. 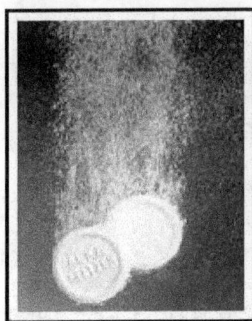 As the water fizzes and bubbles, a spoon is stirred into the glass to mix the contents and it visually melts-away. Of course, the spoon is made of pure gallium and the Alka-Seltzer tablets are for visual effect.

This effect was tested by placing one gram of gallium in an Alka-Seltzer bath for 48 hours and there was no apparent adverse effect on the gallium.

Thoughts:

> This effect can be "dressed-up" by using sugar packets pre-loaded with powdered Alka-Seltzer. The powder is made by crushing the tablets.

In performance, the prankster would claim to have ancient alchemical knowledge that enables him to use common ingredients to instantly dissolve metal.

He would then proceed by pouring a little dab of salt or other dissolvable food spices into the glass of water. The last ingredient placed in the glass is the fake sugar (powered Alka-Seltzer) to cause the phony acid reaction that seemingly melts the spoon.

* The method for loading sugar packets with other substances can be found in chapter 10.

Spiral Soup Spoon

In this effect, a spoon is stirred into a cup of instant noodle soup and when the spoon is removed, it has taken on the twisting shape of the noodles.

Method: This comical effect is based on the *Shrinking Spoon*

effect (chapter 13). The spiral spoon is hidden in the cup of soup from the start and the spoon that is used for stirring is one made of pure gallium. As the spoon melts within the soup, the hidden spiral spoon is withdrawn.

The spiral spoon is made by taking the bowl section of a spare spoon and soldering it onto a compression spring. The spring is cut to a size that will allow the spiral spoon to be hidden within the height of the soup cup (4.5 inches).

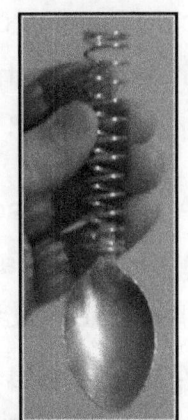

Hot Enough to Fry an Egg or Spoon

On hot summer days, people are fond of saying that you can

fry an egg on a hood of a car. Based on this theme, you can make a prank video using a gallium spoon in place of an egg.

Caution: The spoon is placed on a plate or some holding container and not directly on the hood of the car for obvious reasons.

Cube in Bottle

In theory, a solid metal cube could be "built" inside a narrow

mouth bottle in the same fashion of the classic ship-in-bottle techniques. Once built, the cube's dimension would be larger than the bottle's mouth.

The mold for casting the cube would be a box (an open top square with 5-sides) constructed of cardboard or other thick stock paper. The box would be folded in a manner that would allow it to pass through the mouth of the bottle and then be reshaped within it.

A syringe attached with long plastic tubing would be used to inject liquid gallium or field's metal into the box mold. When the liquid metal solidifies, the mold would be cut away with a long-reaching tool or the bottle would be filled with water to eventually disintegrate the paper.

Chapter 21

Ready-Made Molds

The following ready-made molds are alternatives for the do-it-yourself spoon and ring molds.

* Magnetic Polycarbonate Chocolate Mold

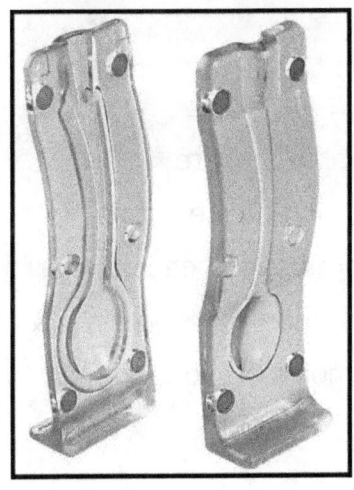

This sturdy 2-piece mold is tightly held together by magnets. The mold was designed to make chocolate spoons.

> Dimensions of spoon: 3.25 inches long and 1 inch wide at the bowl.

> Price: Approximately $20.

The spoon size produced from this mold is slightly over half the size of a regular spoon [6"] but can still be used to mimic an after-dinner coffee spoon or a demitasse spoon.

This mold is perfect for the individual who needs a quick and easy means for casting a gallium spoon for a classroom demonstration.

> *Notes*: The after-dinner coffee spoon is served after a meal with coffee to aid in digestion. It measures in a range of 4½ to 5 inches long--- a length that balances the spoon in a coffee cup.
>
> The demitasse / mocha spoon is approximately 3¾ to 4½ inches long. It is used in formal dining where coffee is served in a demitasse cup and a demitasse spoon is used if sugar is added.

* Antique Bronze Spoon Molds

These slightly rare molds are still available and are occasionally found on auction sites and in antique shops. These types of early century molds were used to cast pewter spoons in a variety of patterns. The common size spoons [7& 3/8 inches long] produced from these molds are longer than today's average spoons of 6 to 6.25 inches.

* Custom Made Spoon Molds

The following on-line sites offer molds design specifically for gallium spoon making:

www.rotometals.com

www.disappearingspoons.com

The suppliers of these molds make one-part types were the spoon is hung in a container while silicone rubber is poured over it, creating a solid block. Once dried, the mold is partially split at the spoon's seam-line.

These molds aren't designed to approximate a generic spoon. The molded spoons have a thin straight-line handle to minimize the amount of gallium needed and to provide a hanging point for the pour-hole and vent-line.

* Ready Made Ring Molds

Off-the-self ring molds can be purchased through mold-making specialists. Even though these molds are limited in ring size and style, you may find one that suits your need.

One of the best sources is:
www.etsy.com

If you google *silicone ring molds*, you can link to other sources as well. These open-top, silicone rubber molds are advertised for crafting a variety of materials, mostly resin, soap, and wax. They are also used to bake food

products up to 395°F. Given this temperature range, the molds are suitable for casting LMP alloys.

Field's metal is an ideal alloy to make a gaffed ring (chapter 16) that will melt-away in a hot liquid. Once the ring is casted, it will be very durable and for all purposes it can be worn as a normal ring. The ring can even be engraved.

Chapter 22

DIY Molds

The Basics of Molding Gallium Gaffs

Materials Needed:

1: Liquid gallium. The metal must first be liquefied in order to pour it into a mold for shaping.

2: Minimum heat source. The gallium container can be set next to an open heat source such as a lamp, until it melts. Given its near-room-temperature melting point (85.57 °F), no flame or expensive chemistry lab equipment is needed.

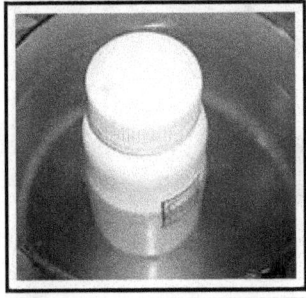

The metal is usually packaged in a leak-proof Kautex plastic bottle. You can place the container in a hot-water bath to liquefy it.

> How hot is tap water? Normally, for residential faucets the maximum temperature of hot water is around 140°F. Many people like to keep the water at about 125°F to prevent scalding. Hot or warm water from a regular faucet is well above the melting point of gallium.

Another option for liquefying the metal is to use a handheld, hair blow-dryer.

It's best to keep your gallium in the original container because it is "pre-stained". Moving the gallium to another container will cause more staining and thus more loss in material.

3: A mold to pour the liquid metal into. The mold is for making the casting of the desired object.

What item you can mold is only limited by the quantity of pure gallium or LMP alloys available in terms of your budget and resources.

Your ability to cast certain mold types for making various gaffs depends on your acquired skill-set and your determination to do so.

4: Amount of gallium required. How many grams that are needed depend on the object being molded.

When casting a spoon mold, a typical size spoon (chapter 7) will create a cavity that will take 20-25 grams of liquid gallium to fill.

Various suppliers of ready-made spoon molds (chapter 21) will suggest a range between 16-20 grams of gallium. This amount is calculated on their specific mold designed with their choice of spoon -- often the thinnest available to control the cost of gallium.

To eliminate costly trial and error attempts at purchasing a sufficient amount of gallium prior to building an actual mold, a minimum of 20 and a maximum of 30 grams are recommended.

The ideal purchase is 25 grams when the amount of gallium can be purchase in multiples of five grams.

5: A Lincoln Syringe (10-12ml), the type used for administering medicines orally. The syringe is used to collect the liquid gallium and then to dispense it into the mold.

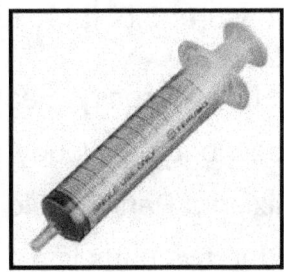

Note: After repeated use for casting, the inner walls of the syringe will become stained with gallium build-up and the tip may become clogged. This situation can be managed by removing the plunger and freeing the solidified portion. If the syringe is the lock-type, submerge it in hot water to liquefy the gallium and then pump it out into a container of cold water to instantly cool it.

Spoon Mold

How to Make a 2-Part Spoon Mold

This spoon mold is based on the generic spoon as detailed in chapter 7. In addition to this information, it is recommendation that you view the various tutorials provided on YouTube and other internet sources to get a visual on the general steps to mold making.

Materials Needed:

A. One spoon to copy.

B. A mold box. Mold boxes can be made of numerous materials such as wood, cardboard, or corrugated plastic sheets. The standard mold box is adjustable for a variety of projects. In this project, the inner dimensions of the box measure: 6.5" [L] X 1.75" [W] X 1.5" [H].

C. Low viscosity, clear liquid silicone. Most clear silicone is described as "cloudy". The silicone is prepared for use by mixing a Part-A liquid with a Part-B liquid. The easiest mixture of the two parts is a 1-to-1 ratio.

D. Wet clay (aka: white modeling clay) or a water based clay.

E. Mold Release. This comes in a spray canister [*Pol-Ease 2500 Release Agent*] and is used to coat the inside walls of the mold box to ensure an easy removal.

Step-1: Lay the clay foundation. Take a lump of clay and flattened it out on a work table. You need enough clay to make a 6" X 8" mat. If preferred, you can use a rolling pin and 2 boards [0.25" thick] to level the clay.

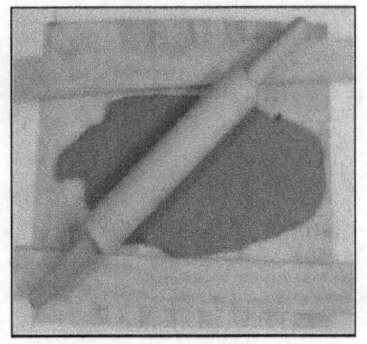

Step-2: Embed the spoon into the clay. The base of the spoon's bowl is pressed into the clay to the point that the handle at the other end doesn't completely submerge into the clay.

Step-3: Shape the clay to the spoon. Because of the spoon's shape, the tip of the bowl will be elevated and the neck area will be slightly gapped.

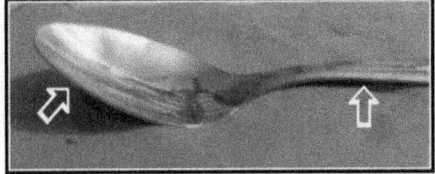

In this situation, small pieces of clay are used to fill-in the elevated spaces to make a solid "bed of clay" for the spoon.

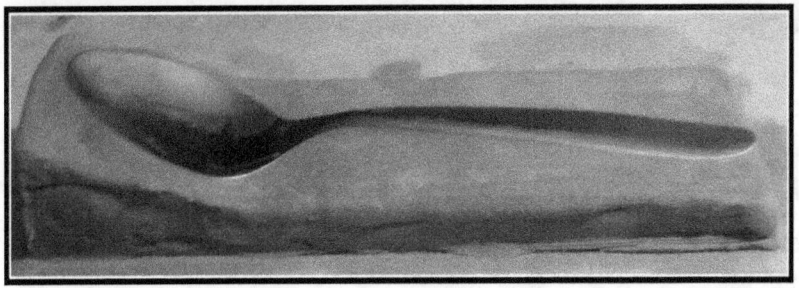

Step-4: Trim the foundation. Take the mold box and center it over the spoon and press down slightly to make a clay outline of the box. Follow the outline and cut away the excess clay.

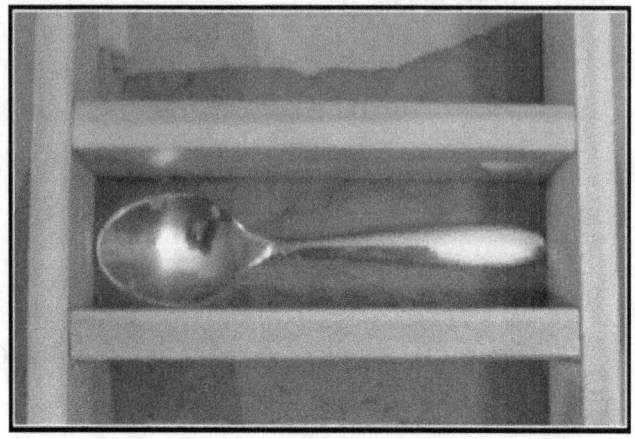

Step-5: Set the keys. Keys are small indentations in the clay that will leave a negative space for the silicone to fill, thus providing locking points to correctly align the two parts of the mold. Keys can be made with small marbles kept in place or the eraser head of a pencil to make the indentations.

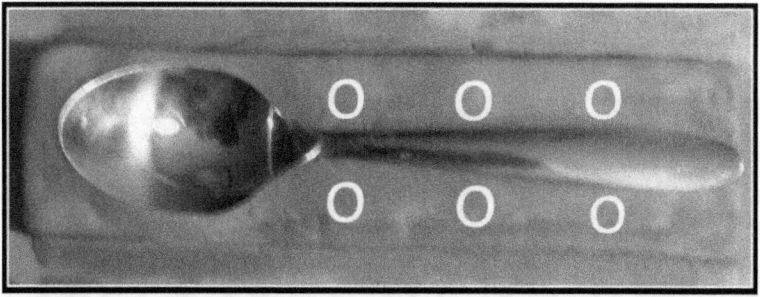

Once done, the box is fitted over the newly shaped foundation.

Step-6: Check the inside of the box to make sure that there are no seam-line gaps between the clay foundation and the inner box. If so, knead some clay to fill any cracks. Clean off any clay residual on the spoon with a damp cloth. Make sure you set the pour-hole and vent-line for the spoon (see end of chapter).

Step-7: Prep the inner walls of the box by spraying on mold-release. This will ensure that the hardened silicone can be easily removed once the mold is formed.

Step-8: Pour the liquid silicone into the box. Start the pouring process in the middle of the box and let it flow outwards. Amount needed for this project: 5 ounces total [Part A: 2.5 & Part B: 2.5] for the first side. Let the silicone dry for 4 to 6 hours.

Step-9: Once dried, turn the mold over and remove the clay. Clean off any residual left on the spoon. Prep the inner walls of the box by spraying on mold-release.

Step-10: For this side of the mold box, pour the liquid silicone as described in Step-8. Let the silicone dry for 4 to 6 hours.

Step-11: Once dried, remove your newly formed mold for casting LMP metals.

Vent Line

In this project, the very tip of the spoon's handle is touching the back-side of the mold box to establish the main "pour hole" for the syringe to inject liquid metals. It's also important to setup a vent-line for air to escape during the casting to ensure an even flow of material.

The vent line is trimmed to size with any object small in circumference, such as a plastic coffee stirrer, paper clip, matchstick or electric wire. The vent line is angled away from main line.

How to Cast a Gallium Spoon

"A cold mold is a no go".

Step-1: Heat the silicone mold to about 100°F for gallium and 150° for Field's metal. By heating the mold, the liquid metal will flow easier, thus preventing any uneven surfaces from developing on the spoon by crystals randomly forming during the pouring process.

Tip: Also heat the syringe when using Field's metal.

The mold can be heated in an oven or by placing it under a

flexible desk lamp [equipped with a 150-Watt, Floodlight] for 20-30 minutes. Do not expose your mold to excess temperatures above 100C / 212°F. Do not microwave the mold.

Step-2: Assemble the 2-part mold and secure it with rubber bands (thin, 4-inch type).

> Wrap one band around the very bottom and one around the top. All bands should be wrapped around the mold 3 times each.

> Wrap one band around the center of the spoon's bowl area and one around the neck area.

> The last two bands are wrapped equal distance on the handle area of the mold.

Too much pressure from over wrapping the bands will restrict the flow of gallium to the lower parts of the mold (bowl area) and too little will cause the gallium to leak from the seams.

You can do a dry-run of the mold by banding it and pouring colored water into the mold to check for leaks or restrictions. Don't use soda because the carbonated bubbles will create misleading restrictions.

> See end of this section for securing the mold with acrylic plates

Step-3: Fill the syringe with liquid metal to about the 2.5ml level for 20-22 grams of gallium. Press the tip of the syringe into the pour-hole on top of the mold and inject the gallium very slowly to avoid creating air pockets. When the mold is properly filled, a drop of gallium will surface to the top of the mold through the vent-hole.

Step-4: Let the gallium solidify. Depending on how hot the mold is, it may take some time for the gallium to harden. You can speed up the process by taking a solid piece of gallium and placing it on the droplet that formed out of the vent-hole.

It's a bad idea to move the mold before the solidification process is complete. If you disrupt a crystal from properly forming, an uneven surface on the spoon will occur.

Step-5: Remove the spoon from the mold. If there are any burrs on the spoon from the gallium expanding through the seam lines, heat a knife in hot water to melt them.

Trouble Shooting

* If the mold doesn't fill properly, do not open it because you'll have a mess to deal with. Place it in the freezer until the gallium hardens and then open the mold to retrieve the contents. Use a mild household cleaner and warm water to remove the staining on the mold.

* Should you spill a good amount of gallium during the casting process, refer to the end of chapter-2 for information on how to handle spillages.

Acrylic Plates to Secure the Mold

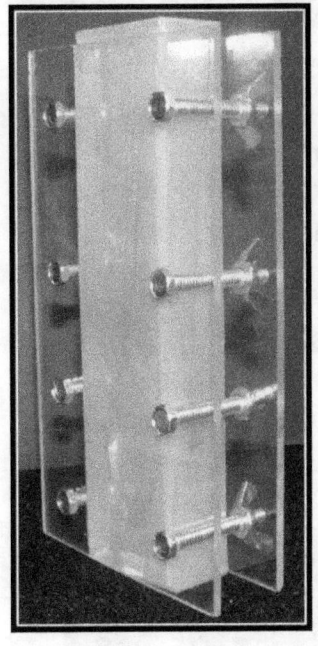

Another option to secure the mold for casting is to utilize acrylic plates. The see-through plates are placed on the front and backside of the mold to form a holding frame around it.

The frame is constructed by drilling matching holes into each plate to allow the positioning of bolts to hold them in place. The mold is sandwiched between to two plates and wing-nuts are used to tighten to the bolts to provide an even pressure to secure the mold.

Open- Top Mold for Coin, Medallion, or Ring

When making a coin mold, the only practical mold that can be crafted by the average hobbyist is an open-top mold.

This type of mold will produce a coin showing only one side (heads or tails) of the coin's profile and the opposite side will be blank. This limitation is adjusted for in a routine (chapter 15) by showing only the detailed side of the gaffed coin while holding it within the plastic bag from the spectator's point of view.

> A *silicone putty mold* is ideal for making molds of the older-style coins that have a raised, high- definition profile.

> Silicone putty comes in two separate components which are kneaded together to make the overall mold. The putty has the consistency of cookie dough when mixed. The object to be molded is pressed into the putty or the putty is shaped around the object to make the mold.

> In order to capture the weak details of the newer, lightly stamped coins, you'll need a low viscosity, *liquid silicone rubber* that will reproduce the finest details.

Here's how to make an easy, open-top mold for pouring liquid silicone:

Step-1: Take a small plastic container and turn it over and cut out the bottom to make an opening. This is the new opening to pour the liquid silicone into.

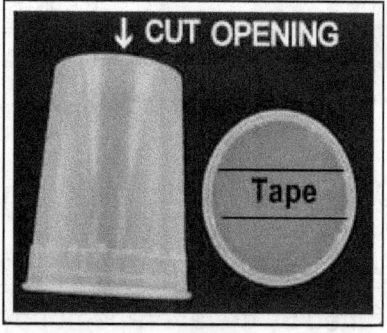

Step-2: Take the lid and turn it over. Next, take a strip of double-sided tape and press it onto the lid's underside.

Step-3: Take the object that you want to copy and press it onto the double-sided tape to hold it in place. Make sure to center the object.

Step-4: Secure the container onto the lid. This upside-down cup and lid is now your container for pouring your mold.

Step-5: Pour the liquid silicone into the container's opening and let the mold sit until the silicone dries.

Step-6: Once dried, remove the lid to extract the new mold from its holding container. The embedded object is detached from the mass of rubber to finish the mold.

Open Mold Casting

Step-1: Heat the mold.

Step-2: Use a syringe to fill the cavity with the liquid metal.

Step-3: If the metal bubbles over the top of the cavity, take the length of a toothpick and scape away the excess.

Step-4: Let the metal solidify and then remove the copy.

Chapter 23

Portable Temp Control Devices

The THERMOS 18 Cube Ice Mat

This reusable ice pack comes in a 9" x 12" size mat that contains 18 connected, liquid filled cubes to provide refrigeration. Each cube is 3" x 2" and the bulk thickness per cube is half an inch.

The cubes can be cut to configure "ice bags" for cold-storing the molded medallion or coin as described in chapter 15.

Small Size Ice-Bag for Jacket Pockets

This bag is made by cutting-out two connected cubes from the mat. The cubes are turned sideways and then sandwiched together by duct taping the ends. Because the two cubes are

squeezed together, the inner sections form a double cushion. The gallium-indium gaff is inserted between the cushions to be held in place until needed.

Small-Size Ice Bag for Pants Pockets

This bag is the thin version that is ideal for pants pockets.

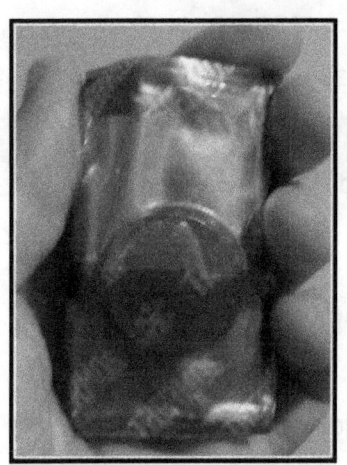

The bag is made by cutting-out two connected cubes from the mat. The cubes are kept in a vertical position and the one on the right side is sliced open to allow the contents to drain off. The cut is made horizontally across the entire mid-section of the cube.

Note: Be sure not to cut through to the backside layer of the pouch.

This deflated pouch is now placed against the backside of the unaltered cube and duct taped into place. Now, you have a single cube and on its backside is a slotted pocket for carrying a gallium-indium gaff. The bag is placed in the pants pocket facing outward. This position enables the thumb and forefinger to easily slide into the slot and retrieve the gaff.

Ice Bag for Spoons

Depending on the size of your spoon or the number of spoons that you want to store at one time, you can cut and configure the shape that you need from the mat.

A spoon ice-bag is for safely transporting your spoon without it prematurely melting. The bag can be placed in a prop box or suitcase until needed.

The bag is mainly designed to store a gallium-indium spoon but may serve the same purpose for one made of pure gallium. Under most circumstance a gallium gaff doesn't need the protection of refrigeration but a performer working outdoors in hot temperatures may need an ice-bag to shield it from the heat until needed.

These ice bags will provide about 90 minutes of refrigerated protection for your gaffs. The overall time can be extended by placing the bags into a mini- ice chest for transportation.

The Blaze Craze Deck

The BCD is a magician's utility device for covertly generating heat inside a poker-size tuck case. The heat is electrically produced with a mini- heat coil wired to three AAA batteries.

The gadget can be used to "vanish" thermal sensitive inks and to reshape objects made of nitinol memory-wire. In chapter 14, it is used to heat a gallium-indium gaff for a quicker melt process.

The DIY hobbyist can make the BCD with the following materials:

* One AAA, three compartment battery holder and batteries.

* 30 feet of winding wire [28 or 30 gauge] or enamel-coated magnet wire.

* Some duct tape, a glue stick and double-sided tape.

* 3 bridged-size playing cards and a poker-size tuck case.

To Make the Coil Unit:

Step-1: Take the three bridge size playing cards and glue them together to make a support board. If you don't have bridge-size cards, take 3 poker-size cards and slightly trim the sides of each card. Another option for the board is a plastic gift card (credit card size).

Step-2: Take 30 feet of winding wire and wind it around the center section of the board to make a coil. This is started by taking one end of the wire and taping it to the board. The wire should overlap the board by an inch. This protruding wire will be #1 of the lead wires.

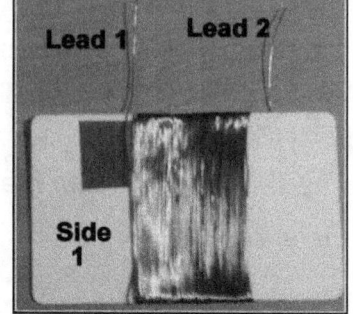

Step-3: Wrap all the wire around the board. The ending section of the wire should extend about an inch beyond the board to make the second lead wire. This wire will end up on side-2 of the board and is taped into place.

Note: The loops of wire are wrapped side-by-side. If the loops are overlapped, layer upon layer, the coil board will be too thick for the tuck case.

Step-4: Take the negative and positive lead wires from the battery-holder and solder them to the lead wires on the coil.

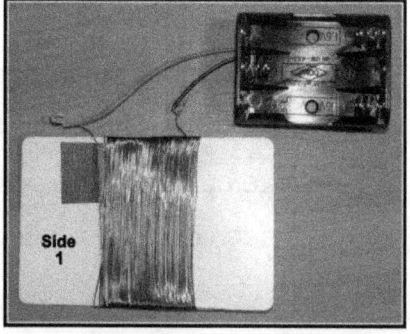

Step-5: Place double-sided tape on the bottom side of the battery holder and then secure it to side-2 of the coil-board.

> The coil-board is inserted into the tuck case with side-1 facing the side of the card box that you want to heat-up. To make a good tight

fit, place one or more playing cards below the batteries. The coil unit will produce 120° of heat by drawing one Amp. The batteries will last 30-45 minutes.

To turn the unit on, install all three batteries, to turn it off, simply remove one of the batteries.

The Mimesis Card Workshop

The Mimesis Workshop (that's just me and my three dogs) makes the rechargeable, electronic version of the *Blaze Craze Deck*.

This version produces enough heat to ignite flash paper or cotton. Yes, flash paper can be ignited without a flame. The unit is built with a push-button, LED lights and a USB port for over 400 charges.

The readers of this book can access the instructional information on the BCD by going to:

> **www.mimesis-magic.com**

> See side menu: Video Center

> Password: blazecraze007.

Appendix A

Supporting Materials

[Items listed in alphabetical order]

Covert Set-up

An important point to know about magic done at the dinner table or in a restaurant setting is that during the time spent chatting and eating, you have all the opportunity to covertly setup your tricks.

Preparing an effect during meal time is possible because real-world misdirection is part of the performing conditions. Looking at the menu, passing the salt, handling the silverware, food being served and waiters approaching the table are situations that help facilitate opportunities for secretly setting-up materials and methods.

Once your set-up is completed, you should wait near the end of the meal period where there is less activity so that you can command the center of attention. Spectators can't witness the magic if their attention is elsewhere. Sometimes "attention direction" is more important than misdirection.

FriXion, the Science Fiction Pen

Without a doubt, the *FriXion Erasable Gel Pen* by *Pilot* is a performer's dream come true. A committee of magicians and mentalist couldn't have designed a better off-the-shelf pen for magic applications.

The pen is "rewritable", which means that you can erase and 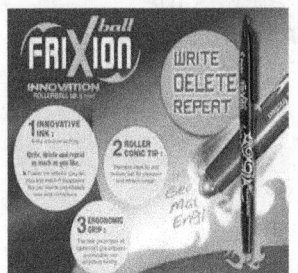 rewrite repeatedly without damaging the paper because the thermo-sensitive ink formula will "disappear" when friction is applied by the pen's eraser. Actually, when heated, the ink becomes clear and thus blends in with the paper's background.

Additionally, the "cleared" ink can be made to reappear by exposing the ink to cold temperatures (14°F), such as in a freezer.

The basic application of the pen is to write words or symbols in permanent, regular black ink and then to disguise the writing with the *FriXion* ink. When heat in the form of a flame is applied to both inks, the *FriXion* ink will disappear and the regular ink will remain in place.

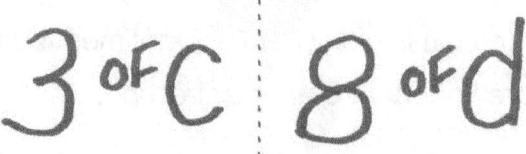

For example, in preparation for a *Wrong-to-Right Prediction* type effect, the *3-of-c* (3 of clubs) would be written on a business card in regular black ink and then the *FiXion* ink would be used to convert the writing into the *8-of-d* (8 of diamonds).

During the performance, a flame from a lighter would be used to heat the back side of the business card and as the spectator concentrates on the writing, it visually morphs from the *8-of-d* into the *3-of-c*.

Spoon Boy Prediction Note

For the *Spoon Boy Prediction* (chapter 14), the entire message and underling [*Concentrate on the Spoon*] is written in *Frixion* black ink except the "S" and the "OO"s which are written in regular black ink. When the message is held over the *Blaze Craze Deck,* the prediction will appear.

As the cliché goes, the effects possible with this pen are only limited by your imagination.

Gas Duster

Gas duster is mistakenly referred to as compressed air or canned air. The product is used for cleaning computer equipment such as keyboards and other electronic devices that cannot be cleaned with water.

A gas duster doesn't use ordinary compressed air as true air dusters do. The canisters actually contain gases that are compressed into liquids, by using chemicals such as 1,1-difluoroethane or, 1,1,1-trifluoroethane.

When gas is highly pressurized [70 psi], a release of pressure will cause a pronounced drop in temperature. The rapid expansion of gas from a pressurized small space into a larger space means that it will absorb heat from the area as it spreads out. During prolong use of a gas duster, the surrounding air and canister will become cold, which is the reason why the contents are labeled with a frostbite warning.

By turning the canister upside down, you're actually spraying forth the chemical contents; a liquid coolant that can cool to as low as -67 °F. This adiabatic cooling property comes in handy when needing to rapidly cool liquid gallium to a solid state.

Prince Rupert's Drops (continued from chapter 15)

The secret to making Prince Rupert's drops started in Mecklenburg, Germany as early as 1625 and were commonly sold as entertainment (toys and party pieces). They were even immortalized in a verse of the *Ballad of Gresham College* (1663).

For those who arrived late and weren't let in on the secret, here's how to make them:

> You start with soft glass, such as soda glass or stirring rods. Do not use glass made of borosilicate (Pyrex). This type of glass can be shaped into a Rupert's drop but it will not explode.

* However, do use borosilicate glass if you want to make a "fake" Rupert's drop that can be given to a spectator as a souvenir.

> The tip of the glass rod is heated with the flame of a meeker  burner or butane torch until a "red hot" molten droplet forms. With steady heat, the droplet will fall free into the receiving container of cold water. As it immediately cools, the toughened glass will form the properties of a Rupert's drop.

* At times, the droplet will shatter upon contact with the water. This is due to the heat being over-applied to the droplet's tail, forcing a rapid separation from the rod. The trick to making a good droplet is to steadily direct the hottest part of the flame

(blue) to the head of the droplet and let the heat transfer gradually free the tail.

Recovery of Liquid Metal

You just finished your trick by using the immersion technique. Your one-time gaffed spoon (or ring), now formless and liquefied, is sitting at the bottom of the coffee cup. How do you recover your liquid metal for further use?

At this point you surely don't want to be perceived as being concerned or attached to the contents of coffee cup. Such a perception would draw attention to the cup's unseen contents.

Here are some suggestions:

* *Time Delay*. Proceed by clearing the table for your next trick and casually set aside any props that you don't want available for inspection. At a later time, you can discreetly secure the recovery of your liquid metal.

* *Slush Containment*. In this method, *Slush Powder* (chapter 10) is used to contain the liquid metal for processing at a later time. When there is no lid for capping the coffee cup, slush powder is poured into the coffee and stirred. Immediately, the powder will gel on top of the liquid metal. This procedure is ideal for transporting your metal without any concerns that it will spill with the coffee. At a later time, the "gel mold" is

removed, turned upside-down and the solidified metal is picked from the gel.

In a performance situation, after an immersion-type trick is completed, the slush powder can be covertly poured into the coffee under the pretense of adding sugar. When the powder firmly gels in place, the performer pretends to drink the coffee while tilting his head back and then turns over the cup (for a few seconds), pattering: "*Good to the last drop*". The perception is that the cup of coffee is now empty. The cup is then placed to the side for the next routine.

In this method, the slush powered is delivered into the coffee with "fake" sugar packets containing the powder (chapter 10) or with a thumb-tip /cork setup.

In the thumb-tip setup, slush powder is loaded into the fake thumb, capped with a cork and placed in the pocket. When the powder is needed, the hand goes into the pocket and the middle-fingers curl around the cork as the thumb pushes it out of the gimmick. The uncorked thumb-tip is worn on the thumb or finger- palmed depending on how much powder has 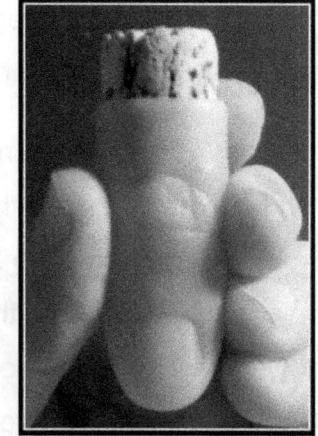 been loaded in it. The thumb-tip is now ready to dump its load.

When the thumb-tip is held in a finger-palmed position, the thumb and index-finger are free to grip objects. This situation enables the performer to pick-up a sugar packet while pretending to tear it open. As the pinched sugar packet and hand turns downward, the slush powder with spill free from the hidden thumb-tip. It looks as if the powder is coming from the sugar packet.

* *Syringe It.* At a discrete moment and place, draw the liquid metal into a syringe. This procedure is done "blind" given that the liquid metal is not visible in the 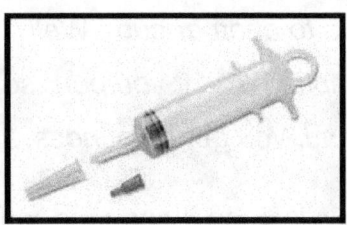 opaqueness of the coffee. It can accurately be removed by slightly tilting the cup downward (towards the body) to localize the bulk of the metal before drawing it into the syringe to be capped. A large, plastic *Thumb Ring Syringe* (60ml) is the perfect tool for this application. The thumb- ring handle allows the plunger to be pulled upward with one hand while the other hand positions the cup.

* *Thumb-Tip Vanish.* In this method, the liquid metal is poured directly into the cavity of the thumb-tip while creating a vanishing.

An XXL thumb-tip will hold 20 grams of liquid gallium with room for a small amount of coffee. See tip at end of this section on how to extend the size of your thumb-tip by using a hollowed out cork.

For the effect, you pour coffee from your cup into another cup and then empty the remaining coffee into your closed fist (with concealed thumb-tip) and follow by making the liquid vanish.

To justify this action, you patter:

"What do you think happens when you pour a liquid from one cup to another cup? Well… one gets filled and the other gets emptied. But if you pour some into the emptiness of your hand, you get emptiness… nothing".

This is done by pouring- off coffee from your cup (containing the liquid metal) into another cup. The pouring process is stopped at the point in which you can visually judge that you have the right amount of coffee remaining in your cup. The correct quantity is the amount that can be poured into the thumb-tip without over-filling it to the point that your physical thumb won't fit.

When the proper amount is determined, the coffee is poured into the thumb-tip and the liquid metal will come along for the ride.

Note: Disposable coffee cups made of paper are the easiest to pour from. These cups can be squeeze to form a "V" shaped spout to control a smooth flow of liquid without dribbling.

Here's how to manage the thumb-tip:

After a pause from the main effect, the thumb-tip is loaded onto the right-thumb. During the trick, the left hand does all the pouring.

At the point in the routine were you have stopped pouring coffee into the receiving cup and you are ready to pour into your fist, your left hand tables the cup.

Then, your right hand (loaded with thumb-tip) points to the left hand as it makes a fist. The right-thumb goes into the hollow of the fist and deposits the thumb-tip. This action is done while you're pattering, *"What happens if you pour something into the emptiness of your hand?"*

Now, the right-hand retrieves the primary cup and pours the contents into the concealed thumb-tip. Once done, the cup is tabled again.

To effect the vanishing, your right forefingers cover the knuckles of your closed left-hand as the right-thumb steals-away the thumb-tip. You dramatically blow on your hands, open them; flex your fingers to show "emptiness". Both hands are then placed on the table with the thumbs overlapping the edge to keep the thumb-tip concealed.

At the first opportunity, the thumb-tip is corked and pocketed for safe-keeping.

> *Tip:* Take a cork and hollow out the center section as much as possible without compromising it. When the thumb-tip is corked, any overflow of liquid will be pushed into the cavity without spillage. In essence, you have extended the carrying capacity of your thumb-tip.

* *Take It.* In a café setting, cap your disposable coffee cup with a lid. When you leave, take it with you. This is not unusual given the price of some coffees.

* *Trash It.* In a café setting, cap your disposable coffee cup with a lid and be seen trashing it. When all is clear, come back and retrieve the cup.

Re-Light-able Matches

This simulated "burnt" match is a realistic, disposable gaff that will give confidence to the performer who values the best in his props.

The matches are made to appear as burnt. They will relight when struck against the striking surface of the matchbox.

Each match has a burnt, flamed stem with charcoal spotting.

This is NOT just a match that has been blackened with a marker. The reduced, small head is used to approximate the size of a burnt match.

Materials Needed: *Strike-On-Box* Wooden Matches (32 Count).

How to Make Re-light-able Matches

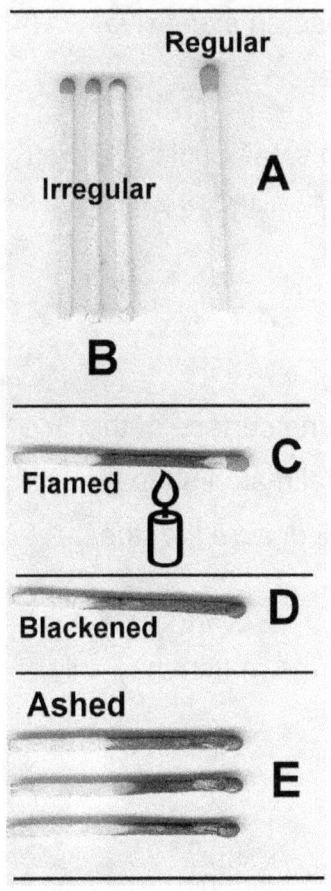

A & B. Select the Irregular Matches

If you search through a box of matches, you'll find some very irregular matches that have small heads --- these are the ones that you want to collect and alter.

C. Flame the Match

Take a match and hold it over a candle flame to light the middle section of the stick while rotating it. The trick is to not char the stick too deep and to ensure that the flame doesn't ignite the head. This procedure will give the match a true burnt look.

D. Blacken the Match

Take a black marker and blacken the unburnt portion up to the head and then blacken the head completely. The blackening should be put on to make a complete covering but shouldn't be overdone because excessive coats of ink will cause the match to smolder when struck on the matchbox.

E. Ashen the Match

Take some acrylic, silver paint and dab on the smallest amount possible to give the match an ash, charcoal look. The paint should be applied sparingly because an excessive amount will cause the match to be useless.

This type of paint can be purchased in a small bottle at a craft or hobby store. The brand that I use is: *FolkArt- PLAID, 506 Silver*, 2 ounce bottle.

How to use:

Strike the gaffed match against the striking surface of the matchbox, as you would do with a normal match. Grip the match at the center of the stem. This will ensure that the match doesn't break in half.

Do not hesitate when striking. If you do, the match may only smolder.

Other notes:

> The simulated burnt matches should be kept in the same box as the regular matches. The contrast between the matches (burnt & unburnt) aids in the visual deception of the gaff. This set-up also leads into the introductory patter of:

"*You know* [open matchbox] *some people throw away the burnt matches* [remove "burnt" match] *but I like to keep them, they're as good as the un-burnt ones* [display both matches together], *you just need to reactivate the flame by blowing hot air onto the burnt ones* [blow on the gaff match and strike it on the box]. *There you go, a penny saved is a penny earned.*"

> If the gaffed match breaks in half when you try to strike it, you can exploit the moment by taking the remainder of the match and re-striking it. This situation becomes even more dramatic.

> Should the match smolder when attempting to light it, you can capitalize on the situation by pattering: *"Ok, the burnt match almost lit… but someone here is a non-believer when it comes to things magical. Let's try another match and this time everyone needs to truly believe"!*

Temperature Teaching Tool

The *Zoo Med Repti-Temp Digital Infrared Thermometer* is an inexpensive, pocket-size [1.25 X 2.5 inches] device for measuring surface temperatures.

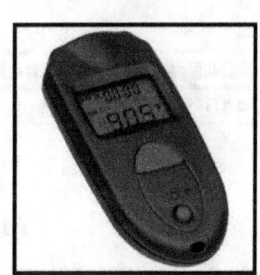

Beyond its normal function, the thermometer's small size is ideal for inconspicuously checking room temperatures and surfaces prior to a performance in an unfamiliar setting.

Appendix B

Material Safety Data Sheet
Gallium, 99.99%

Section 1 - Chemical Product and Company Identification

MSDS Name: Gallium, 99.99%
Synonyms: None.
For emergencies in the US, call CHEMTREC: 800-424-9300

Section 2 - Composition, Information on Ingredients

CAS#	Chemical Name	Percent	EINECS/ELINCS
7440-55-3	Gallium	99.99	231-163-8

Section 3 - Hazards Identification

EMERGENCY OVERVIEW

Appearance: grey solid.
Warning! Causes respiratory tract irritation. Causes eye and skin irritation. Moisture sensitive. Corrosive to metal. The toxicological properties of this material have not been fully investigated.
Target Organs: Blood, eyes, skin, mucous membranes.

Potential Health Effects
Eye: Causes eye irritation.
Skin: Causes skin irritation. May be harmful if absorbed through the skin. The toxicological properties of this material have not been fully investigated. May cause contact dermatitis.

Ingestion: May cause gastrointestinal irritation with nausea, vomiting and diarrhea. The toxicological properties of this substance have not been fully investigated. May be harmful if swallowed.
Inhalation: Causes respiratory tract irritation. The toxicological properties of this substance have not been fully investigated. May be harmful if inhaled.
Chronic: May cause bone marrow abnormalities with damage to blood forming tissues. Administration of gallium to humans has caused metallic taste, skin rashes, and bone marrow depression.

Section 4 - First Aid Measures

Eyes: In case of contact, immediately flush eyes with plenty of water for a t least 15 minutes. Get medical aid.
Skin: In case of contact, flush skin with plenty of water. Remove contaminated clothing and shoes. Get medical aid if irritation develops and persists. Wash clothing before reuse.
Ingestion: If swallowed, do not induce vomiting unless directed to do so by medical personnel. Never give anything by mouth to an unconscious person. Get medical aid.
Inhalation: If inhaled, remove to fresh air. If not breathing, give artificial respiration. If breathing is difficult, give oxygen. Get medical aid.
Notes to Physician: Treat symptomatically and supportively.

Section 5 - Fire Fighting Measures

General Information: As in any fire, wear a self-contained breathing apparatus in pressure-demand, MSHA/NIOSH (approved or equivalent), and full protective gear.
Extinguishing Media: Do NOT use water, CO2, or halogenated extinguishers. Use dry chemical extinguishing agents, dry sand or dry ground dolomite.
Flash Point: Not applicable.
Autoignition Temperature: Not available.
Explosion Limits, Lower: Not available.
Upper: Not available.
NFPA Rating: (estimated) Health: 2; Flammability: 0; Instability: 0

Section 6 - Accidental Release Measures

General Information: Use proper personal protective equipment as indicated in Section 8.
Spills/Leaks: Vacuum or sweep up material and place into a suitable disposal container. Avoid generating dusty conditions. Provide ventilation. Do not get water inside containers. Cool material below 25°C to solidify before attempting cleanup. Protect metal construction as gallium will dissolve the metal/is corrosive to most metals.

Section 7 - Handling and Storage

Handling: Wash thoroughly after handling. Avoid contact with eyes, skin, and clothing. Keep container tightly closed. Avoid ingestion and inhalation. Use with adequate ventilation. Wash clothing before reuse. Keep from contact with moist air and steam.
Storage: Store in a tightly closed container. Store in a cool, dry, well-ventilated area away from incompatible substances. Store protected from moisture.

Section 8 - Exposure Controls, Personal Protection

Engineering Controls: Facilities storing or utilizing this material should be equipped with an eyewash facility and a safety shower. Use adequate ventilation to keep airborne concentrations low.
Exposure Limits

Chemical Name	ACGIH	NIOSH	OSHA - Final PELs
Gallium	none listed	none listed	none listed

OSHA Vacated PELs: Gallium: No OSHA Vacated PELs are listed for this chemical.
Personal Protective Equipment
Eyes: Wear appropriate protective eyeglasses or chemical safety goggles as described by OSHA's eye and face protection regulations in 29 CFR 1910.133 or European Standard EN166.
Skin: Wear appropriate protective gloves to prevent skin exposure.
Clothing: Wear appropriate protective clothing to prevent skin exposure.
Respirators: A respiratory protection program that meets OSHA's 29 CFR 1910.134 and ANSI Z88.2 requirements or

European Standard EN 149 must be followed whenever workplace conditions warrant respirator use.

Section 9 - Physical and Chemical Properties

Physical State: Solid
Appearance: grey
Odor: Not available.
pH: Not available.
Vapor Pressure: 1 mm Hg @ 1350 deg C
Vapor Density: Not available.
Evaporation Rate: Not available.
Viscosity: Not available.
Boiling Point: 2400 deg C
Freezing/Melting Point: 29 deg C
Decomposition Temperature: Not available.
Solubility: Insoluble.
Specific Gravity/Density: 5.9
Molecular Formula: Ga
Molecular Weight: 69.72

Section 10 - Stability and Reactivity

Chemical Stability: Gallium is stable in dry air. It tarnishes in moist air or oxygen.
Conditions to Avoid: Dust generation, moisture, exposure to air, metals, exposure to moist air or water.
Incompatibilities with Other Materials: Hydrogen peroxides, hydrochloric acid, halogens, phosphorus, sulfur, alkalies, oxygen, metals, oxidizing agents.
Hazardous Decomposition Products: Exposure to moist air or water.
Hazardous Polymerization: Has not been reported

Section 11 - Additional Information

... and a partridge in a pear tree.

ABOUT THE AUTHOR

The author is a Summa Cum Laude graduate of psychology and a life-long student of magic. He has created over **67** marketed magic effects and currently operates the website: www.mimesis-magic.com.

The author is known as the *coin guy* who "splits" cards and is the sole craftsman who makes all the gaffed & gimmick devices for the popular
Mimesis Card Workshop

"May your deceptions dare to dazzle and delight"

www.ingramcontent.com/pod-product-compliance
Lightning Source LLC
Chambersburg PA
CBHW051653170526
45167CB00001B/443